Functional Verification of Programmable Embedded Architectures

A Top-Down Approach

FUNCTIONAL VERIFICATION OF PROGRAMMABLE EMBEDDED ARCHITECTURES

A Top-Down Approach

PRABHAT MISHRA
Department of Computer and Information Science and Engineering
University of Florida, USA

NIKIL D. DUTT
Center for Embedded Computer Systems
Donald Bren School of Information and Computer Sciences
University of California, Irvine, USA

Prabhat Mishra
University of Florida
USA

Nikil D. Dutt
University of California, Irvine
USA

Functional Verification of Programmable Embedded Architectures
A Top-Down Approach

ISBN- 1-4899-7336-2
ISBN- 978-1-4899-7336-8
ISBN 0-387-26399-3 Printed on acid-free paper.

Softcover re-print of the Hardcover 1st edition 2005

9 8 7 6 5 4 3 2 1 SPIN 11430100

springeronline.com

To our families.

Contents

Preface xv

Acknowledgments xix

I Introduction to Functional Verification 1

1 Introduction 3
- 1.1 Motivation . . . 3
  - 1.1.1 Growth of Design Complexity . . . 3
  - 1.1.2 Functional Verification - A Challenge . . . 4
- 1.2 Traditional Validation Flow . . . 8
- 1.3 Top-Down Validation Methodology . . . 10
- 1.4 Book Organization . . . 12

II Architecture Specification 13

2 Architecture Specification 15
- 2.1 Architecture Description Languages . . . 16
  - 2.1.1 Behavioral ADLs . . . 18
  - 2.1.2 Structural ADLs . . . 19
  - 2.1.3 Mixed ADLs . . . 19
  - 2.1.4 Partial ADLs . . . 20
- 2.2 ADLs and Other Specification Languages . . . 20
- 2.3 Specification using EXPRESSION ADL . . . 21
  - 2.3.1 Processor Specification . . . 24
  - 2.3.2 Coprocessor Specification . . . 25
  - 2.3.3 Memory Subsystem Specification . . . 27
- 2.4 Chapter Summary . . . 28

3 **Validation of Specification** **29**
3.1 Validation of Static Behavior 30
3.1.1 Graph-based Modeling of Pipelines 31
3.1.2 Validation of Pipeline Specifications 34
3.1.3 Experiments 45
3.2 Validation of Dynamic Behavior 48
3.2.1 FSM-based Modeling of Processor Pipelines 48
3.2.2 Validation of Dynamic Properties 54
3.2.3 A Case Study 59
3.3 Related Work 61
3.4 Chapter Summary 62

III Top-Down Validation **63**

4 **Executable Model Generation** **65**
4.1 Survey of Contemporary Architectures 66
4.1.1 Summary of Architectures Studied 66
4.1.2 Similarities and Differences 68
4.2 Functional Abstraction 69
4.2.1 Structure of a Generic Processor 69
4.2.2 Behavior of a Generic Processor 73
4.2.3 Structure of a Generic Memory Subsystem 74
4.2.4 Generic Controller 74
4.2.5 Interrupts and Exceptions 75
4.3 Reference Model Generation 77
4.4 Related Work 80
4.5 Chapter Summary 81

5 **Design Validation** **83**
5.1 Property Checking using Symbolic Simulation 85
5.2 Equivalence Checking 87
5.3 Experiments 88
5.3.1 Property Checking of a Memory Management Unit 88
5.3.2 Equivalence Checking of the DLX Architecture 91
5.4 Related Work 92
5.5 Chapter Summary 93

6 Functional Test Generation **95**
6.1 Test Generation using Model Checking 95
6.1.1 Test Generation Methodology 96
6.1.2 A Case Study 99
6.2 Functional Coverage driven Test Generation 103
6.2.1 Functional Fault Models 103
6.2.2 Functional Coverage Estimation 105
6.2.3 Test Generation Techniques 106
6.2.4 A Case Study 112
6.3 Related Work 116
6.4 Chapter Summary 117

IV Future Directions **119**

7 Conclusions **121**
7.1 Research Contributions 121
7.2 Future Directions 122

V Appendices **125**

A Survey of Contemporary ADLs **127**
A.1 Structural ADLs 127
A.2 Behavioral ADLs 130
A.3 Mixed ADLs 134
A.4 Partial ADLs 139

B Specification of DLX Processor **141**

C Interrupts & Exceptions in ADL **147**

D Validation of DLX Specification **151**

E Design Space Exploration **155**
E.1 Simulator Generation and Exploration 156
E.2 Hardware Generation and Exploration 162

References **167**

Index **179**

List of Figures

1.1 An example embedded system 4
1.2 Exponential growth of number of transistors per integrated circuit 5
1.3 North America re-spin statistics 6
1.4 Complexity matters 7
1.5 Pre-silicon logic bugs per generation 8
1.6 Traditional validation flow 9
1.7 Proposed specification-driven validation methodology 11

2.1 ADL-driven exploration and validation of programmable architectures 16
2.2 Taxonomy of ADLs 17
2.3 Commonality between ADLs and non-ADLs 21
2.4 Block level description of an example architecture 22
2.5 Pipeline level description of the DLX processor shown in Figure 2.4 23
2.6 Specification of the processor structure using EXPRESSION ADL 24
2.7 Specification of the processor behavior using EXPRESSION ADL 25
2.8 Coprocessor specification using EXPRESSION ADL 26
2.9 Memory subsystem specification using EXPRESSION ADL . . . 27

3.1 Validation of pipeline specifications 30
3.2 An example architecture 32
3.3 A fragment of the behavior graph 33
3.4 An example processor with false pipeline paths 36
3.5 An example processor with false data-transfer paths 37
3.6 The DLX architecture 46
3.7 ADL driven validation of pipeline specifications 49
3.8 A fragment of a processor pipeline 50
3.9 The processor pipeline with only instruction registers 51
3.10 Automatic validation framework using SMV 59

3.11 Automatic validation framework using equation solver 60

4.1 A fetch unit example 70
4.2 Modeling of RenameRegister function using sub-functions 72
4.3 Modeling of MAC operation 73
4.4 Modeling of associative cache function using sub-functions 74
4.5 Example of distributed control 75
4.6 Example of centralized control 76
4.7 Mapping between *MACcc* and generic instructions 78
4.8 Simulation model generation for the DLX architecture 79

5.1 Top-down validation methodology 84
5.2 Test vectors for validation of an *AND* gate 85
5.3 Compare point matching between reference and implementation design 87
5.4 TLB block diagram 89

6.1 Test program generation methodology 97
6.2 A fragment of the DLX architecture 100
6.3 Test Generation and Coverage Estimation 112
6.4 Validation of the Implementation 114

C.1 Specification of division_by_zero exception 148
C.2 Specification of illegal_slot_instruction exception 148
C.3 Specification of machine_reset exception 149
C.4 Specification of interrupts 149

D.1 The DLX processor with pipeline registers 152

E.1 Architecture exploration framework 156
E.2 Cycle counts for different graduation styles 158
E.3 Functional unit versus coprocessor 160
E.4 Cycle counts for the memory configurations 162
E.5 The application program 163
E.6 Pipeline path exploration 164
E.7 Pipeline stage exploration 165
E.8 Instruction-set exploration 166

List of Tables

3.1 Specification validation time for different architectures 45
3.2 Summary of property violations during DSE 48
3.3 Validation of in-order execution by two frameworks 61

4.1 Processor-memory features of different architectures. *R4K: MIPS R4000, SA: StrongArm, 56K: Motorola 56K, c5x: TI C5x, c6x: TI C6x, MA: MAP1000A, SC: Starcore, R10: MIPS R10000, MP: Motorola MPC7450, U3: SUN UltraSparc IIi, α64: Alpha 21364, IA64: Intel IA-64* 67
4.2 A list of common sub-functions 71

5.1 Validation of the DLX implementation using equivalence checking 91

6.1 Number of test programs in different categories 99
6.2 Reduced number of test programs 100
6.3 Test programs for validation of DLX architecture 115
6.4 Quality of the proposed functional fault model 115
6.5 Test programs for validation of LEON2 processor 116

E.1 The Memory Subsystem Configurations 161
E.2 Synthesis Results: RISC-DLX vs Public-DLX 162

Preface

It is widely acknowledged that the cost of validation and testing comprises a significant percentage of the overall development costs for electronic systems today, and is expected to escalate sharply in the future. Many studies have shown that up to 70% of the design development time and resources are spent on functional verification. Functional errors manifest themselves very early in the design flow, and unless they are detected up front, they can result in severe consequences – both financially and from a safety viewpoint. Indeed, several recent instances of high-profile functional errors (e.g., the Pentium FDIV bug) have resulted in increased attention paid to verifying the functional correctness of designs. Recent efforts have proposed augmenting the traditional RTL simulation-based validation methodology with formal techniques in an attempt to uncover hard-to-find corner cases, with the goal of trying to reach RTL functional verification closure. However, what is often not highlighted is the fact that in spite of the tremendous time and effort put into such efforts at the RTL and lower levels of abstraction, the complexity of contemporary embedded systems makes it difficult to guarantee functional correctness at the system level under all possible operational scenarios.

The problem is exacerbated in current System-on-Chip (SOC) design methodologies that employ Intellectual Property (IP) blocks composed of processor cores, coprocessors, and memory subsystems. Functional verification becomes one of the major bottlenecks in the design of such systems. A critical challenge in the validation of such systems is the lack of an initial golden reference model against which implementations can be verified through the various phases of design refinement, implementation changes, as well as changes in the functional specification itself. As a result, many existing validation techniques employ a bottom-up approach to design verification, where the functionality of an existing architecture is, in essence, reverse-engineered from its implementation. For instance, a functional model of an embedded processor is extracted from its RTL implementation, and this functional model is then validated in an attempt to verify the functional correctness of the implemented RTL.

If an initial golden reference model is available, it can be used to generate reference models at lower levels of abstraction, against which design implementations can be compared. This "ideal" flow would allow for a consistent set of reference models to be maintained, through various iterations of specification changes, design refinement, and implementation changes. Unfortunately such golden reference models are not available in practice, and thus traditional validation techniques employ different reference models depending on the abstraction level and verification task (e.g., functional simulation or property checking), resulting in potential inconsistencies between multiple reference models.

In this book we present a top-down validation methodology for programmable embedded architectures that complements the existing bottom-up approaches. Our methodology leverages the system architect's knowledge about the behavior of the design through an architecture specification that serves as the initial golden reference model. Of course, the model itself should be validated to ensure that it conforms to the architect's intended behavior; we present validation techniques to ensure that the static and dynamic behaviors of the specified architecture are well formed. The validated specification is then used as a golden reference model for the ensuing phases of the design.

Traditionally, a major challenge in a top-down validation methodology is the ability to generate executable models from the specification for a wide variety of programmable architectures. We have developed a functional abstraction technique that enables specification-driven generation of executable models such as a simulator and synthesizable hardware. The generated simulator and hardware models are used for functional validation and design space exploration of programmable architectures.

This book addresses two fundamental challenges in functional verification: lack of a golden reference model, and lack of a comprehensive functional coverage metric. First, the top-down validation methodology uses the generated hardware as a reference model to verify the hand-written implementation using a combination of symbolic simulation and equivalence checking. Second, we have proposed a functional coverage metric and the attendant task of coverage-driven test generation for validation of pipelined processors. The experiments demonstrate the utility of the specification-driven validation methodology for programmable architectures.

We begin in Chapter 1 by highlighting the challenges in functional verification of programmable architectures, and relating a traditional bottom-up validation approach against our proposed top-down validation methodology. Chapter 2 introduces the notion of an Architecture Description Language (ADL) that can be used as a golden reference model for validation and exploration of programmable architectures. We survey contemporary ADLs and analyze the features required in ADLs to enable concise descriptions of the wide variety of programmable architectures. We also describe the role of ADLs in generating software tools and hardware models from the specification.

In Chapter 3, we present techniques to validate the ADL specification. In the context of pipelined programmable architectures, we describe methods to verify both static and dynamic behaviors embodied in the ADL, with the goal of ensuring that the architecture specified in the ADL conforms to the system designer's intent, and is consistent and well-formed with respect to the desired architectural properties.

Chapter 4 focuses on the important notion of functional abstraction that permits the extraction of key parameters from the wide range of contemporary programmable architectures. Using this functional abstraction technique, we show how various reference models can be generated for the downstream tasks of compilation, simulation and hardware synthesis. In Chapter 5, we show how the generated hardware models can be used to verify the correctness of the hand-written RTL implementation using a combination of symbolic simulation and equivalence checking.

Chapter 6 introduces the notion of functional fault models and coverage estimation techniques for validation of pipelined programmable architectures. We present specification-driven functional test-generation techniques based on the functional coverage metrics described in the chapter. Finally, Chapter 7 concludes the book with a short discussion of future research directions.

Audience

This book is designed for graduate students, researchers, CAD tool developers, designers, and managers interested in the development of tools, techniques and methodologies for system-level design, microprocessor validation, design space exploration and functional verification of embedded systems.

About the Authors

Prabhat Mishra is an Assistant Professor in the Department of Computer and Information Science and Engineering at the University of Florida. He received his B.E. from Jadavpur University, India, M.Tech. from Indian Institute of Technology, Kharagpur, and Ph.D from University of California, Irvine – all in Computer Science. He worked in various semiconductor and design automation companies including Intel, Motorola, Texas Instruments and Synopsys. He received the Outstanding Dissertation Award from the European Design Automation Association in 2005 and the CODES+ISSS Best Paper Award in 2003. He has published more than 25 papers in the embedded systems field. His research interests include design and verification of embedded systems, reconfigurable computing, VLSI CAD, and computer architecture.

Nikil Dutt is a Professor in the Donald Bren School of Information and Computer Sciences at the University of California, Irvine. He received a Ph.D. in Computer Science from the University of Illinois at Urbana-Champaign in 1989. He has been an active researcher in design automation and embedded systems since 1986, with four books, more than 200 publications and several best paper awards. Currently, he serves as Editor-in-Chief of ACM TODAES and as Associate Editor of ACM TECS. He has served on the steering, organizing, and program committees of several premier CAD and embedded system related conferences and workshops. He serves on the advisory boards of ACM SIGBED and ACM SIGDA, and is Vice-Chair of IFIP WG 10.5. His research interests include embedded systems design automation, computer architecture, optimizing compilers, system specification techniques, and distributed embedded systems.

Acknowledgments

This book is the result of many years of academic research work and industrial collaborations. We would like to acknowledge our sponsors for providing us the opportunity to perform the research. This work was partially supported by NSF (CCR-0203813, CCR-0205712, MIP-9708067), DARPA (F33615-00-C-1632), Motorola Inc. and Hitachi Ltd.

This book has the footprints of many collaborations. We would like to acknowledge the contributions of Dr. Magdy Abadir, Jonas Astrom, Dr. Peter Grun, Ashok Halambi, Arun Kejariwal, Dr. Narayanan Krishnamurthy, Dr. Mahesh Mamidipaka, Prof. Alex Nicolau, Dr. Frederic Rousseau, Prof. Sandeep Shukla, and Prof. Hiroyuki Tomiyama. We are also thankful to all the members of the ACES laboratory at the Center for Embedded Computer Systems for interesting discussions and fruitful collaborations.

Acknowledgments

[illegible]

Part I

Introduction to Functional Verification

1

INTRODUCTION

1.1 Motivation

Computing is an integral part of daily life. We encounter two types of computing devices everyday: desktop based computing devices and embedded systems. Desktop based systems encompass traditional computers including personal computers, notebook computers, workstations and servers. Embedded systems are ubiquitous: they run the computing devices hidden inside a vast array of everyday products and appliances such as cell phones, toys, handheld PDAs, cameras, and microwave ovens. Both types of computing devices use programmable components such as processors, coprocessors and memories to execute the application programs. In this book, we refer these programmable components as *programmable embedded architectures* (*programmable architectures* in short). Figure 1.1 shows an example embedded system that contains programmable components as well as application specific hardwares, interfaces, controllers and peripherals.

1.1.1 Growth of Design Complexity

The complexity of the programmable architectures is increasing at an exponential rate. There are two factors that contribute to this complexity growth: technology and demand. First, there is an exponential growth in the number of transistors per integrated circuit, as characterized by Moore's Law [32]. Figure 1.2 shows that Intel processors followed the Moore's law in terms of doubling transistors in every couple of years. This trend is not limited to only high-end general purpose microprocessors. Exponential growth in design complexity is also present in application specific embedded systems. For example, Figure 1.2 also shows the dramatic increase of design complexity for various system-on-chip (SOC) architectures in last few years.

The technology has enabled an exponential increase in computational capacity, which fuels the second trend: the realization of ever more complex applications in the domains of communication, multimedia, networking, and entertainment. For example, the volume of Internet traffic (data movement) is growing exponentially. This would require increase in computation power to manipulate the data. The need for computational complexity further fuels the technological advancement in terms of design complexity.

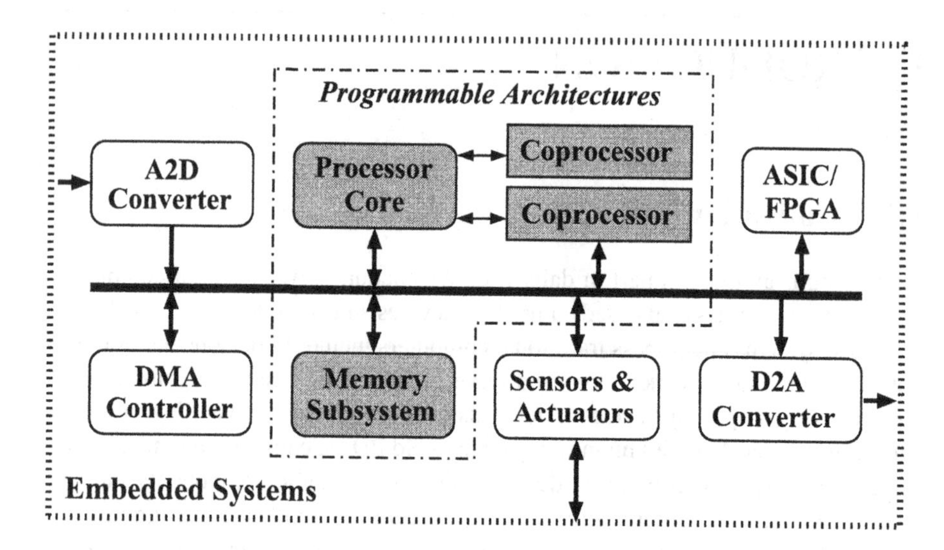

Figure 1.1: An example embedded system

However, the complexity of designing and verifying such systems is also increasing at an exponential rate. Figure 1.3 shows a recent study on the number of first silicon re-spins of system-on-chip (SOC) designs in North America [33]. Almost half of the designs fail the very first time. This failure has tremendous impact on cost for two reasons. First, the delay in getting the working silicon drastically reduces the market share. Second, the manufacturing (fabrication) cost is extremely high. The same study also concluded that 71% of SOC re-spins are due to logic bugs (functional errors).

1.1.2 Functional Verification - A Challenge

Functional verification is widely acknowledged as a major bottleneck in design methodology: up to 70% of the design development time and resources are spent on functional verification [119]. Recent study highlights the challenges of functional verification: Figure 1.4 shows the statistics of the SOC designs in terms of design complexity (logic gates), design time (engineer years), and verification complexity (simulation vectors) [33]. The study highlights the tremendous complexity faced by simulation-based validation of complex SOCs: it estimates that by 2007, a complex SOC will need 2000 engineer years to write 25 million lines of register-transfer level (RTL) code and one trillion simulation vectors for functional verification.

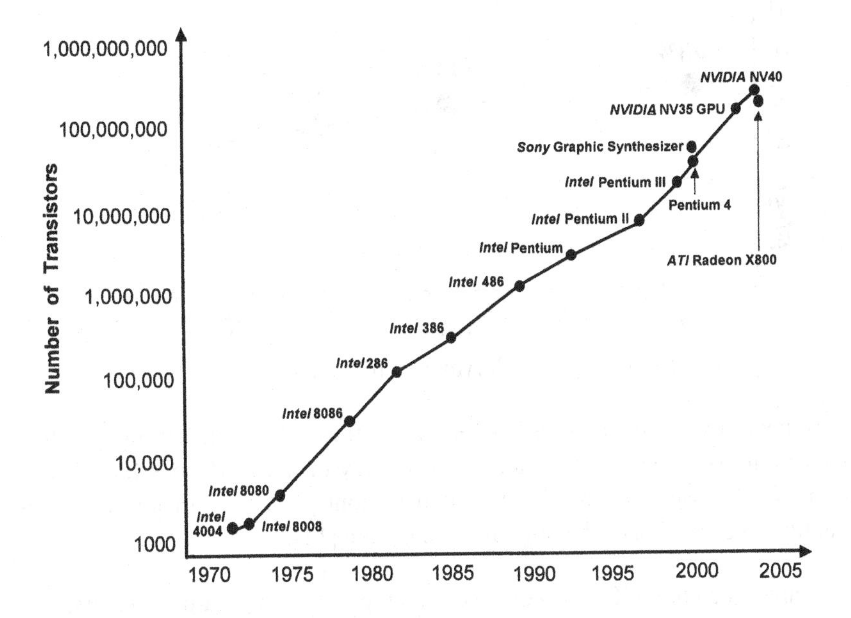

Figure 1.2: Exponential growth of number of transistors per integrated circuit

A similar trend can be observed in the high-performance microprocessor space. Figure 1.5 summarizes a study of the pre-silicon logic bugs found in the Intel IA32 family of microarchitectures. This trend again shows an exponential increase in the number of logic bugs: a growth rate of 300-400% from one generation to the next. The bug rate is linearly proportional to the number of lines of structural RTL code in each design, indicating a roughly constant density [11].

Simple extrapolation indicates that unless a radically new approach is employed, we can expect to see 20-30K bugs designed into the next generation and 100K in the subsequent generation. Clearly – in the face of shrinking time-to-markets – the amount of validation effort rapidly becomes intractable, and will significantly impact product schedules, with the additional risk of shipping products with undetected bugs.

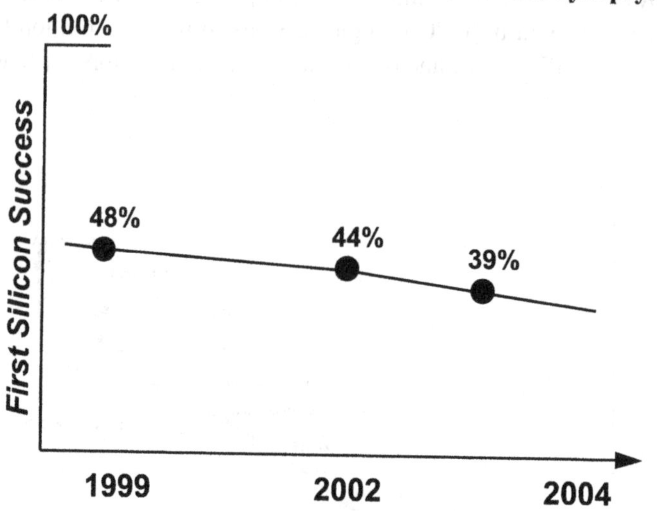

Figure 1.3: North America re-spin statistics

The next obvious question is - where do all these bugs come from? An Intel report summarized the results of a statistical study of the 7855 bugs found in the Pentium 4 processor design prior to initial tapeout [11]. The major categories, amounting to over 75% of the bugs analyzed, were [11]:

- Careless coding (12.7%) - this includes typos and cut-and-paste errors.
- Miscommunication (11.4%) - these errors are due to communication gap.
- Microarchitecture (9.3%) - flaws or omissions in the definition.
- Logic/Microcode changes (9.3%) - errors due to design changes to fix bugs.
- Corner cases (8%)
- Power down issues (5.7%) - errors due to extensive clock gating features.
- Documentation (4.4%) - bugs due to incorrect/incomplete documentation.
- Complexity (3.9%) - bugs specifically due to microarchitectural complexity.
- Random initialization (3.4%) - bugs due to incorrect state initialization.

- Late definition (2.8%) - bugs due to late addition of new features.
- Incorrect RTL assertions (2.8%)
- Design mistake (2.6%) - incorrect implementation errors.

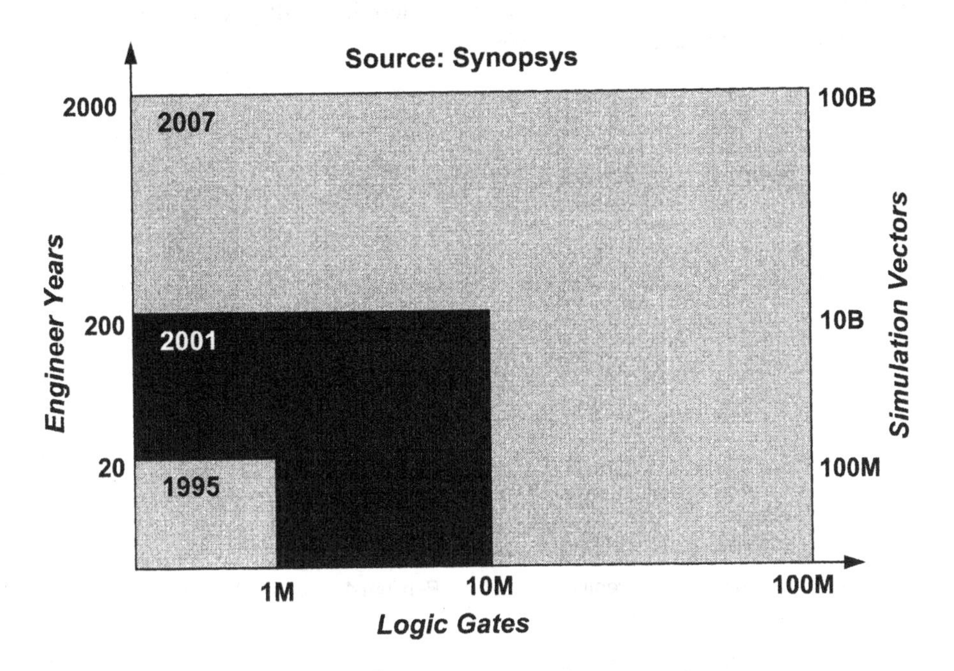

Figure 1.4: Complexity matters

Although "complexity" is ranked eighth on the list of bug causes, it is clear that it contributes to many of the categories listed above. More complex microarchitectures need more extensive documentation to describe them; they require larger design teams to implement them, increasing the likelihood of miscommunication between team members; and they introduce more corner cases, resulting in undiscovered bugs. Hence, microarchitectural complexity is the major contributor of the logic bugs.

Typically, there are two fundamental reasons for so many logic bugs: lack of a golden reference model and lack of a comprehensive functional coverage metric. First, there are multiple specification models above the RTL level (functional model, timing model, verification model, etc.). The consistency of these models is a major concern due to lack of a golden reference model. Second, the design verification problem is further aggravated due to lack of a functional coverage metric that can be used to determine the coverage of the microarchitectural features, as well as the quality of functional validation. Several coverage measures are commonly used during design validation, such as code coverage, finite-state machine (FSM) coverage, and so on. Unfortunately, these measures do not have any direct relationship to the functionality of the design. For example, in the case of a pipelined processor, none of these measures determine if all possible interactions of hazards, stalls and exceptions are verified.

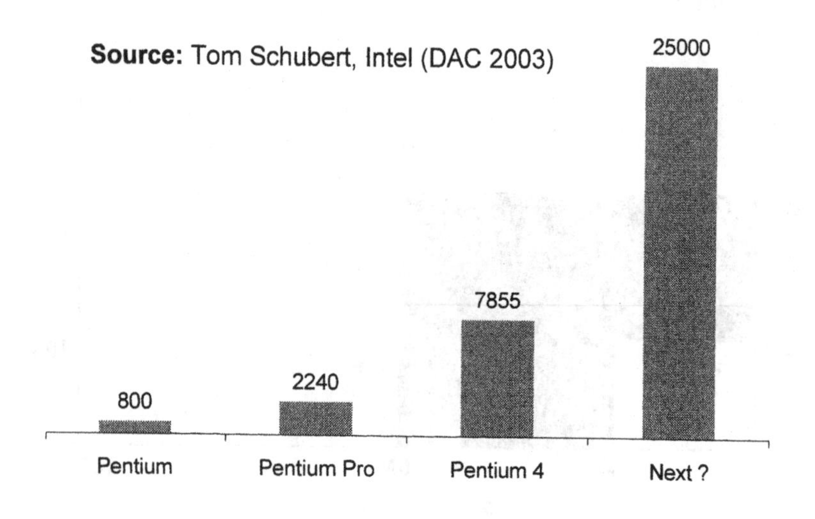

Figure 1.5: Pre-silicon logic bugs per generation

This book presents a top-down validation methodology that addresses the two fundamental challenges mentioned above. We apply this methodology to verify programmable architectures consisting of a processor core, coprocessors, and memory subsystem [110].

1.2 Traditional Validation Flow

Figure 1.6 shows a traditional architecture validation flow. In the current validation methodology, the architect prepares an informal specification of the programmable architectures in the form of an English document. The logic designer implements the modules at the register-transfer level (RTL). The validation effort tries to uncover two types of faults: architectural flaws and implementation bugs. Validation is performed at different levels of abstraction to capture these faults. For example, architecture-level modeling (*HLM* in Figure 1.6) and instruction-set simulation is used to estimate performance as well as verify the functional behavior of the architecture. A combination of simulation techniques and formal methods are used to uncover implementation bugs in the *RTL design*.

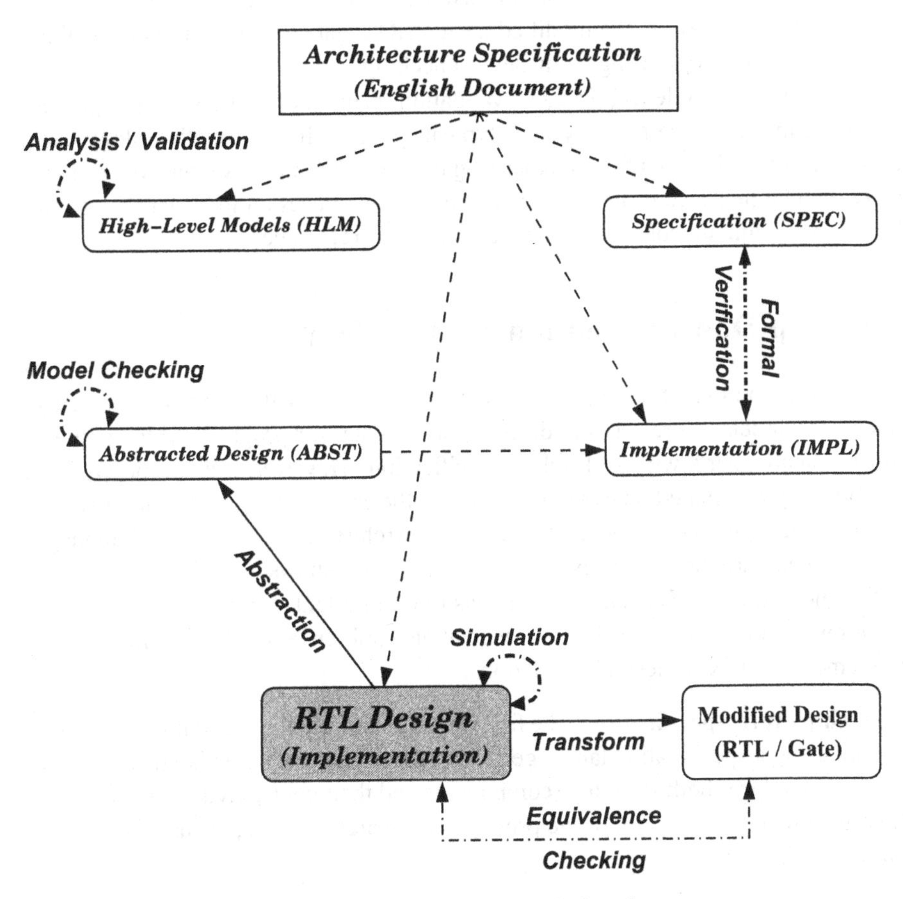

Figure 1.6: Traditional validation flow

Simulation using random (or directed-random) testcases [1, 19, 37, 63, 123] is the most widely used form of microprocessor validation. It is not possible to apply formal techniques directly on million-gate designs. For example, model checking is typically applied on the high-level description of the design (*ABST* in Figure 1.6) abstracted from the RTL implementation [90, 115]. Traditional formal verification is performed by describing the system using a formal language [53, 20, 118, 62, 64, 78, 79]. The specification (*SPEC* in Figure 1.6) for the formal verification is derived from the architecture description. The implementation (*IMPL* in Figure 1.6) for the formal verification can be derived either from the architecture specification or from the abstracted design. In current practice, the validated *RTL design* is used as a golden reference model for future design modifications. For example, when design transformations (including synthesis) are applied on the *RTL design*, the modified design (RTL/gate level) is validated against the golden *RTL design* using equivalence checking.

A significant bottleneck in these validation techniques is the lack of a golden reference model above RTL level. A typical design validation methodology contains multiple reference models depending on the abstraction level and verification activity. The presence of multiple reference models raises an important question: how do we maintain consistency between so many reference models?

1.3 Top-Down Validation Methodology

We propose the use of a single specification to automatically generate necessary reference models. Currently the design methodology for programmable architectures typically starts with an English specification. However, it is not possible to perform any automated analysis or model synthesis on a design specified using a natural language. We propose the use of an Architecture Description Language (ADL) to capture the design specification. Figure 1.7 shows our ADL-driven validation methodology. The methodology has four important steps: architecture specification, validation of specification, executable (reference) model generation, and implementation (RTL design) validation.

1. Architecture Specification: The first step is to capture the programmable architecture using a specification language. Although any specification language can be used that captures both structure (components and their connectivity) and behavior (instruction-set description) of the programmable architectures, we use an ADL in our methodology.

2. Validation of Specification: The next step is to verify the specification to ensure the correctness of the architecture specified. We have developed validation techniques to ensure that the architectural specification is well formed by analyzing both static and dynamic behaviors of the specified architecture. We present algorithms to verify several architectural properties, such as connectedness, false pipeline and data-transfer paths, completeness, and finiteness [99]. The dynamic behavior is verified by analyzing the instruction flow in the pipeline using a FSM based model to validate several important architectural properties such as determinism and in-order execution in the presence of hazards and multiple exceptions

[105, 96]. The validated ADL specification is used as a golden reference model to generate various executable models such as simulator and hardware implementation.

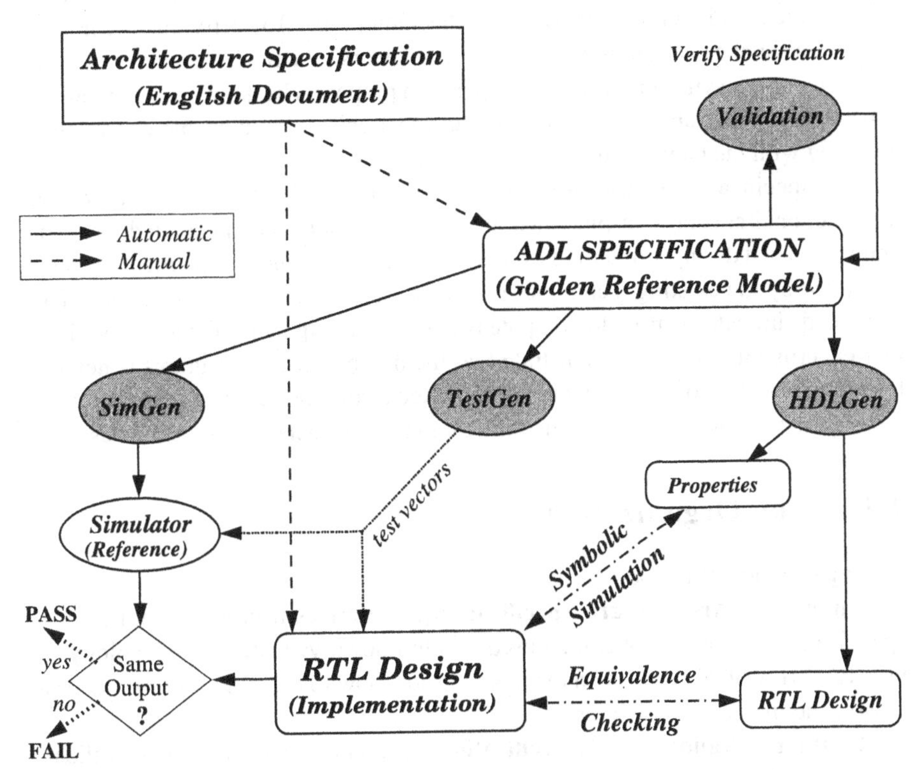

Figure 1.7: Proposed specification-driven validation methodology

3. Executable Model Generation: A major challenge in a top-down validation methodology is the ability to generate executable models from the specification for a wide variety of programmable architectures including RISC (Reduced Instruction Set Computer), DSP (Digital Signal Processor), VLIW (Very Long Instruction Word), and superscalar architectures. We have developed a functional abstraction approach by studying the similarities and differences of each architectural feature in various architecture domains. Based on our observations we have defined necessary generic functions, sub-functions, and computational environment needed to capture a wide variety of programmable architectures. Our functional abstraction technique enables generation of simulators [102], hardware prototypes [93], and models for property checking [100] from the ADL specification.

4. RTL Design Validation: This book explores two validation scenarios using the generated models: design validation using equivalence checking and test generation for functional validation. The generated hardware is used as a reference model for verifying the hand-written implementation (RTL Design) using a combination of symbolic simulation and equivalence checking [106]. To verify that the implementation satisfies certain properties, our framework generates the intended properties and uses a symbolic simulator to perform property checking. Our framework generates synthesizable RTL description of the architecture to enable equivalence checking with the hand-written implementation.

The specification is also used to generate functional test programs based on the functional coverage of pipelined architectures [100]. The generated test programs are used during simulation of the implementation, and complement the tests generated by the existing techniques such as a random test pattern generator. The generated simulator is used to compute the expected outputs for the test programs. Our experimental results demonstrate that the number of test programs generated by our approach to obtain a functional coverage is an order of magnitude less than those generated by random or constrained-random test generation techniques.

1.4 Book Organization

The organization of the book is as follows.

Chapter 2 [Architecture Specification]: Describes necessary features of a specification language that can be used in a top-down validation methodology. It uses EXPRESSION ADL as an example to show how to capture processor, coprocessor, and memory architectures.

Chapter 3 [Validation of Specification]: Presents the techniques for validating the architecture specification. These techniques verify both static and dynamic behaviors of the specified architecture.

Chapter 4 [Model Generation]: Describes automatic generation of models for simulation, hardware generation, and validation for a wide variety of programmable architectures.

Chapter 5 [Design Validation]: The generated hardware is used as a reference model for verifying the hand-written RTL implementation using a combination of symbolic simulation and equivalence checking.

Chapter 6 [Test Generation]: Presents specification-driven test generation techniques based on functional coverage of the pipelined processor architectures.

Chapter 7 [Conclusions]: Contains a summary of the book and a discussion of future research directions.

Part II

Architecture Specification

2

ARCHITECTURE SPECIFICATION

The first step in a top-down validation methodology is to capture the programmable architectures using a specification language. The language should be powerful enough to specify the wide spectrum of contemporary processor, coprocessor, and memory features. On the other hand, the language should be simple enough to allow correlation of the information between the specification and the architecture manual. Specifications widely in use today are still written informally in natural language like English. Since natural language specifications are not amenable to automated analysis, there are possibilities of ambiguity, incompleteness, and contradiction: all problems that can lead to different interpretations of the specification.

Many formal and semi-formal specification languages for describing software and hardware designs have been proposed over the years. The languages range in expressiveness, and their different levels of granularity determine their appropriateness for different applications. This chapter analyzes several types of specification languages and evaluates the suitability of Architecture Description Languages (ADL) in specifying programmable architectures. We use EXPRESSION ADL [5] to illustrate architecture specification using examples. Within this context, it is important to note that this book does not propose a new language or endorse an existing one. The validation techniques presented in this book can use any existing language that captures both structure and behavior of the programmable architectures.

This chapter is organized as follows. Section 2.1 introduces the notion of an architecture description language and surveys the existing ADLs in terms of their specification capabilities. Section 2.2 analyzes different types of languages and evaluates their suitability in specifying programmable architectures. Section 2.3 describes architecture specification using EXPRESSION ADL [5]. Finally, Section 2.4 summarizes the chapter.

2.1 Architecture Description Languages

The phrase "Architecture Description Language" (ADL) has been used in context of designing both software and hardware architectures. Software ADLs are used for representing and analyzing software architectures ([83], [109]). They capture the behavioral specifications of the components and their interactions that comprises the software architecture. However, hardware ADLs capture the structure (hardware components and their connectivity), and the behavior (instruction-set) of processor architectures. In this book the term ADL will refer to hardware architecture description languages.

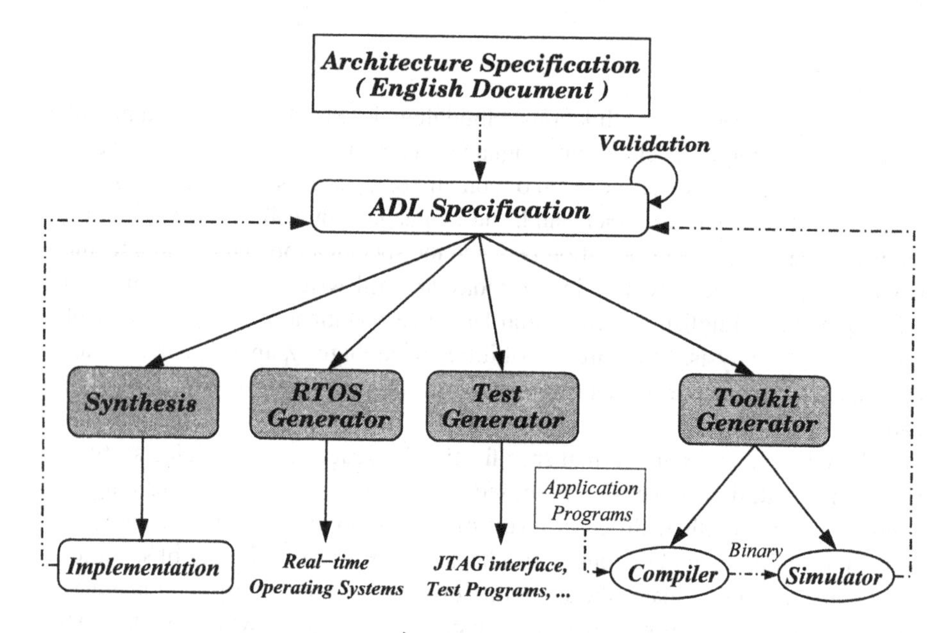

Figure 2.1: ADL-driven exploration and validation of programmable architectures

As embedded systems become ubiquitous, there is an urgent need to facilitate rapid design space exploration (DSE) of programmable architectures. ADLs are used to perform early exploration, synthesis, test generation, and validation of processor-based designs as shown in Figure 2.1. Programmable architectures are captured using an ADL. The ADL specification can be used for generation of a software toolkit including the compiler, assembler, simulator, and debugger. The application programs are compiled and simulated, and the feedback is used to modify the ADL specification with the goal of finding the best possible architecture for the given set of applications. The ADL specification can also be used for generating hardware prototypes under design constraints such as area, power, and clock speed. Several researchers have shown the usefulness of ADL-driven generation of functional test programs and test interfaces. The specification can also be used to generate device drivers for real-time operating systems [124].

Although, ADL-driven exploration is extensively used in both academia (nML [72], ISDL [31], EXPRESSION [5], Valen-C [6], MIMOLA [117], Sim-nML [132], and LISA [133]), and industry (ARC [10], Axys [42], RADL [15], Target [49], Tensilica [130], LISATek [17], and MDES [80]), to the best of our knowledge, there has not been any effort in validating the ADL specification. It is necessary to validate the ADL specification of the architecture to ensure the correctness of both the architecture specified, as well as the generated software toolkit. Chapter 3 presents specification validation techniques for programmable architectures.

Figure 2.2 shows the classification of architecture description languages (ADLs) based on two aspects: *content* and *objective*. The content-oriented classification is based on the nature of the information an ADL can capture, whereas the objective-oriented classification is based on the purpose of an ADL. Contemporary ADLs can be classified into six categories based on the objective: simulation-oriented, synthesis-oriented, test-oriented, compilation-oriented, validation-oriented, and operating system (OS) oriented.

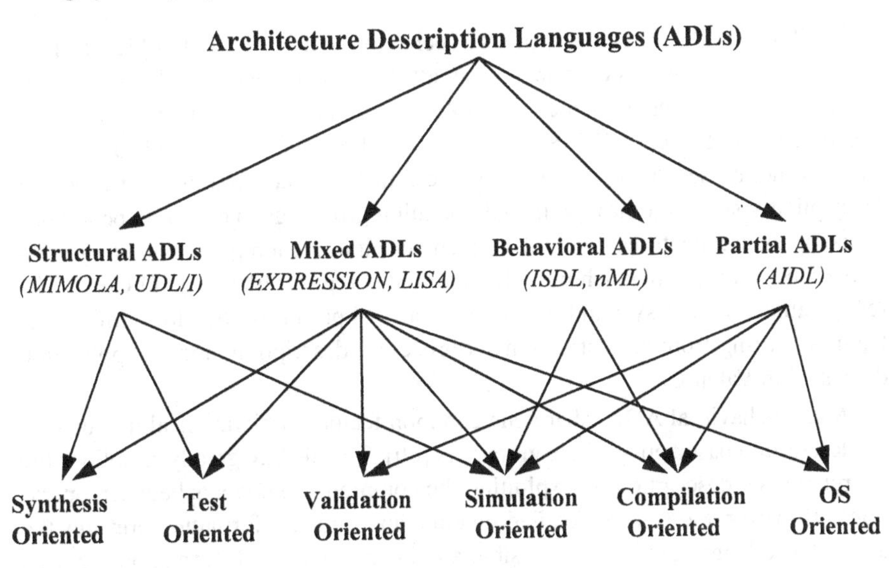

Figure 2.2: Taxonomy of ADLs

ADLs can be classified into four categories based on the nature of the information: structural, behavioral, mixed, and partial. The structural ADLs capture the structure in terms of architectural components and their connectivity. The behavioral ADLs capture the instruction-set behavior of the processor architecture. The mixed ADLs capture both structure and behavior of the architecture. These ADLs capture complete description of the structure or behavior or both. However, the partial ADLs capture specific information about the architecture for the intended task. For example, an ADL intended for interface synthesis does not require internal structure or behavior of the processor.

Traditionally, structural ADLs are suitable for synthesis and test-generation. Similarly, behavioral ADLs are suitable for simulation and compilation. It is not always possible to establish a one-to-one correspondence between content-based and objective-based classification. For example, depending on the nature and amount of information captured, partial ADLs can represent any one or more classes of the objective-based ADLs. Recently, many ADLs have been proposed that capture both the structure and the behavior of the architecture. This section presents a brief survey using content-based classification of ADLs. A detailed survey is available in Appendix A.

2.1.1 Behavioral ADLs

nML [72] and ISDL [31] are examples of behavior-centric ADLs. In nML, the processor's instruction-set is described as an attributed grammar with the derivations reflecting the set of legal instructions. nML has been used by the retargetable code generation environment CHESS [22] to describe DSP and ASIP (Application Specific Instruction-set Processor) architectures. In ISDL, constraints on parallelism are explicitly specified through illegal operation groupings. This could be tedious for complex architectures like DSPs which permit operation parallelism (e.g. Motorola 56K) and VLIW machines with distributed register files (e.g. TI C6X). The retargetable compiler system by Yasuura et al. [6] produces code for RISC architectures starting from an instruction-set processor description, and an application described in Valen-C.

Many behavioral ADLs share one common feature: a hierarchical instruction-set description based on attribute grammars [60]. This feature greatly simplifies the instruction-set description by exploiting the common components between operations. However, the lack of detailed pipeline and timing information prevents the use of these languages as an extensible architecture model. Information required by resource-based scheduling algorithms cannot be obtained directly from the description. Also, it is impossible to generate cycle accurate simulators based on the behavioral descriptions without some assumptions on the architecture's control behavior, i.e., an implied architecture template has to be used.

2.1.2 Structural ADLs

MIMOLA [117] and UDL/I [34] are examples of ADLs that primarily capture the structure of the processor: the net-list of the target processor is described in a HDL (Hardware Description Language) like language. One advantage of this approach is that the same description is used for both processor synthesis and code generation. The target processor has a micro-coded architecture. In MIMOLA, the net-list description is used to extract the instruction-set [116, 117], and produce the code generator. UDL/I [34] is used for describing processors at an RT-level on a per-cycle basis. The instruction-set is automatically extracted from the UDL/I description [35], and is then used for generation of a compiler and a simulator.

In general, structural ADLs enable flexible and precise micro-architecture descriptions. The same description can be used for hardware synthesis, test generation, simulation and compilation. However, it is difficult to extract the instruction-set for retargetable compilation.

2.1.3 Mixed ADLs

More recently, languages that capture both the structure and the behavior of the processor, as well as detailed pipeline information have been proposed (EXPRESSION [5], RADL [15], FLEXWARE [108], MDes [80], and LISA [133]). The main characteristic of LISA is the operation-level description of the pipeline. RADL [15] is an extension of the LISA approach that focuses on explicit support of detailed pipeline behavior to enable generation of production quality cycle-accurate and phase-accurate simulators. FLEXWARE [108] and MDes [80] have a mixed-level structural/behavioral representation. FLEXWARE contains the CODESYN code-generator and the Insulin simulator for ASIPs. The simulator uses a VHDL model of a generic parameterizable machine. The application is translated from the user-defined target instruction-set to the instruction-set of this generic machine. The MDes [80] language used in the Trimaran system is a mixed-level ADL, intended for exploration of parameterized VLIW architectures. Information is broken down into sections (such as format, resource-usage, latency, operation, and register), based on a high-level classification of the information being represented.

The EXPRESSION ADL also follows a mixed-level approach to facilitate DSE. Furthermore, it provides support for specification of novel memory subsystems. It avoids explicit representation of the reservation tables[1] by extracting them from the structural description [88]. The ADL is used to drive the generation of both compiler [4] and simulator [8].

[1]Reservation Tables (RTs) have been used to detect conflicts between instructions that simultaneously access the same architectural resource.

2.1.4 Partial ADLs

The ADLs discussed so far captures complete description of the processor's structure, behavior or both. There are many ADLs that captures partial information of the architecture needed to perform specific task. For example, AIDL aims at validation of pipeline behavior such as data-forwarding and out-of-order completion. AIDL is an ADL developed at University of Tsukuba for design of high-performance superscalar processors [129]. In AIDL, timing behavior of pipeline is described using interval temporal logic. AIDL does not support software toolkit generation. However, AIDL descriptions can be simulated using the AIDL simulator.

2.2 ADLs and Other Specification Languages

There are various types of specification languages including ADLs, programming languages, hardware description languages, modeling languages, and so on. A natural question is whether an ADL is more suitable for specification of programmable architectures than other languages. In other words, how do ADLs differ from non-ADLs? This section attempts to answer this question. However, it is not always possible to answer the following question: given a language for describing an architecture, what are the criteria for deciding whether it is an ADL or not?

In principle, ADLs differ from programming languages because the latter bind all architectural abstractions to specific point solutions whereas ADLs intentionally suppress or vary such binding. In practice, the architecture is embodied and recoverable from code by reverse engineering methods. For example, it might be possible to analyze a piece of code written in C language and figure out if it corresponds to a *Fetch* unit. Many languages provide architecture level views of the system. For example, C++ language offers the ability to describe the structure of a processor by instantiating objects for the components of the architecture. However, C++ offers little or no architecture-level analytical capabilities. Therefore, it is difficult to describe the architecture at a level of abstraction suitable for early analysis and exploration. More importantly, traditional programming languages are not natural choice for describing architectures due to their inability for capturing hardware features such as parallelism and synchronization.

ADLs differ from modeling languages (such as UML) because the latter are more concerned with the behaviors of the whole rather than the parts, whereas ADLs concentrate on representation of components. In practice, many modeling languages allow the representation of cooperating components and can represent architectures reasonably well. However, the lack of an abstraction would make it harder to describe the instruction-set of the architecture.

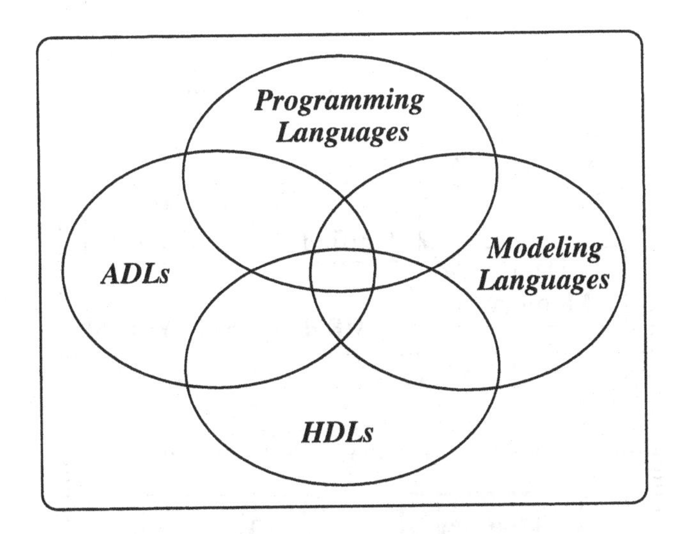

Figure 2.3: Commonality between ADLs and non-ADLs

Traditional hardware description languages (HDL), such as VHDL and Verilog, do not have sufficient abstraction to describe architectures and explore them at the system level. It is possible to perform reverse-engineering to extract the structure of the architecture from the HDL description. However, it is hard to extract the instruction-set behavior of the architecture. In practice, some variants of HDLs work reasonably well as ADLs for specific classes of programmable architectures.

There is no clear line between ADLs and non-ADLs. In principle, programming languages, modeling languages, and hardware description languages have aspects in common with ADLs, as shown in Figure 2.3. Languages can, however, be discriminated from one another according to how much architectural information they can capture and analyze. Languages that were born as ADLs show a clear advantage in this area over languages built for some other purpose and later co-opted to represent architectures.

2.3 Specification using EXPRESSION ADL

Our validation framework uses the EXPRESSION ADL [5] to specify processor, coprocessor, and memory architectures. The EXPRESSION ADL follows a mixed-level approach to facilitate specification of a wide range of programmable architectures. We illustrate the use of the EXPRESSION ADL to describe a simple multi-issue architecture consisting of a processor, a coprocessor and a memory subsystem.

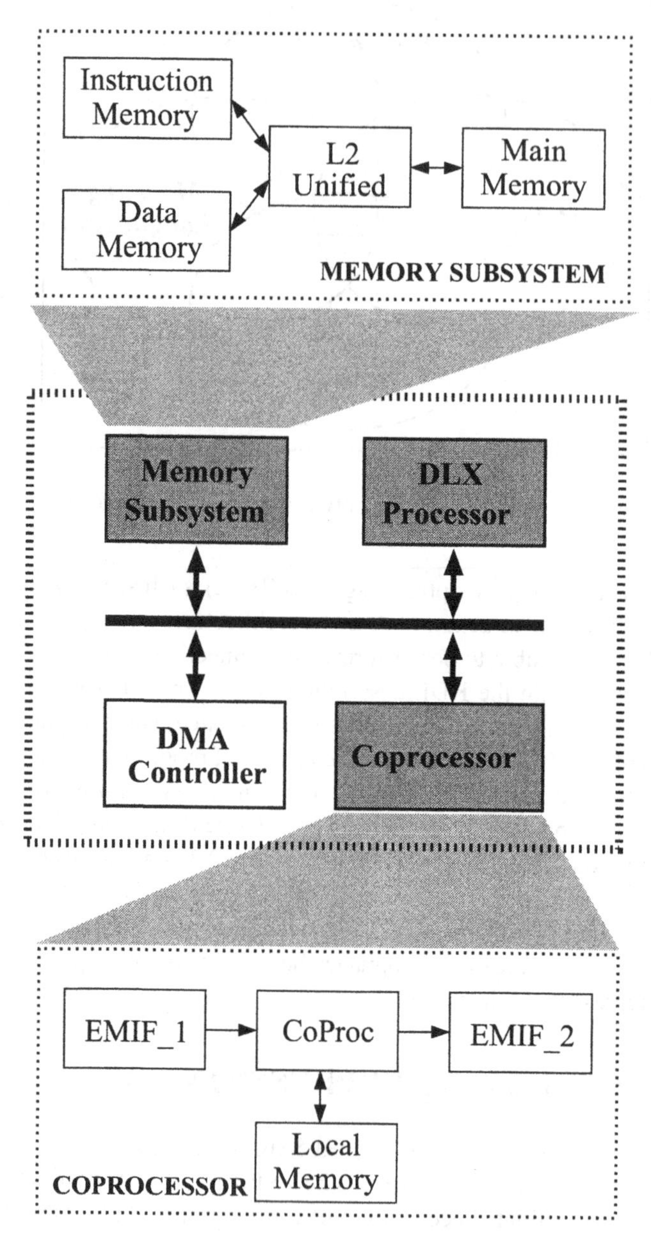

Figure 2.4: Block level description of an example architecture

Figure 2.4 shows the block level description of a simple architecture. This level of detail is available in a typical architecture manual. Typically, pipeline level details are available in a micro-architecture manual. For example, Figure 2.4 shows the detailed description of the memory subsystem and the coprocessor. The memory subsystem consists of separate instruction and data memories (L1 cache), a unified L2 memory, and a main memory. The coprocessor consists of three pipeline stages: EMIF_1 (external memory interface), CoProc, and EMIF_2. The coprocessor uses it local memory for computations. The data transfer between coprocessor local memory and the main memory is handled by the DMA (direct memory access) controller shown in Figure 2.4. Similarly, Figure 2.5 shows the pipeline level description of the DLX processor shown in Figure 2.4. The DLX processor has five pipeline stages: fetch (IF), decode (ID), execute (EX), memory (MEM), and write back (WB). We have chosen the DLX processor since it has been well studied in academia and there are RTL implementations available that can be used for validation.

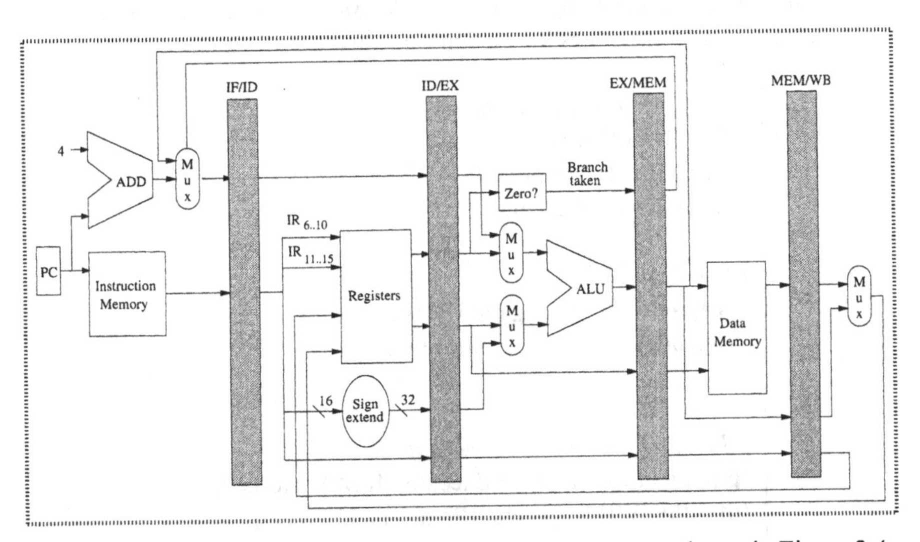

Figure 2.5: Pipeline level description of the DLX processor shown in Figure 2.4

The architecture shown in Figure 2.4 can issue up to two operations (an ALU or memory access operation and a coprocessor operation) per cycle. The coprocessor supports vector arithmetic operations. This section briefly describes how to specify the processor, coprocessor, and memory architectures using the EXPRESSION ADL. The detailed ADL specification of the DLX architecture is available in Appendix B.

2.3.1 Processor Specification

This section describes how the EXPRESSION ADL captures the structure and behavior of the DLX processor shown in Figure 2.5.

Structure

The structure of a processor can be viewed as a net-list with the components as nodes and the connectivity as edges. Figure 2.6 shows a portion of the EXPRESSION description of the processor. It describes all the components in the structure: *PC*, *registers*, *fetch*, *decode*, *ALU*, *MEM*, and *writeback*. Each component has a list of attributes. For example, the *ALU* unit has information regarding the number of instructions executed per cycle, timing of each instruction, supported opcodes, and so on. The connectivity is established using the description of pipeline and data-transfer paths. Informally, a pipeline path is used to transfer instruction whereas a data-transfer path is used to transfer data. For example, $\{IF \rightarrow ID \rightarrow EX \rightarrow MEM \rightarrow WB\}$ is a pipeline path, and $\{WB \rightarrow Registers\}$ is a data-transfer path in Figure 2.5. Section 3.1.1 defines the pipeline and data-transfer paths in detail.

```
# Components specification
( FetchUnit Fetch
  (capacity 2) (timing (all 1))
  (opcodes all) (latches ...) ...
)
( ExecUnit ALU
  (capacity 1) (timing (add 1) (sub 1) ...)
  (opcodes (add sub ...)) (latches ...) ...
)
......
# Pipeline and data-transfer paths
(pipeline Fetch Decode Execute MEM WriteBack)
(dtpaths (WB Registers) (Registers ALU) ...)
......
```

Figure 2.6: Specification of the processor structure using EXPRESSION ADL

Figure 2.6 describes the five-stage pipeline as {*fetch, decode, execute, memory, writeback*}. In this particular case, the execute stage has only one component. In general, the execute stage can have multiple execution paths. Furthermore, each path can contain pipelined or multi-cycle execution units. The ADL specification also includes the description of all the data-transfer paths.

Behavior

The EXPRESSION ADL captures the behavior of the architecture as the description of the instruction-set. The behavior is organized into operation groups, with each group containing a set of operations[2] having some common characteristics. For example, Figure 2.7 shows two operation groups. The *aluOps* includes all the operations supported by the *ALU* unit. Similarly, the *memOps* group contains all the operations supported by the *MEM* unit. Each instruction is then described in terms of its opcode, operands, behavior, and instruction format. Each operand is classified either as source or as destination. Furthermore, each operand is associated with a type that describes the type and size of the data it contains. The instruction format describes the fields of the instruction in both binary and assembly. Figure 2.7 shows the description of the *add* and *store* operations.

```
# Behavior: description of instruction-set
( opgroup aluOps (add, sub, ...) )
( opgroup memOps (load, store, ...) )
......
( opcode add
  (operands (s1 reg) (s2 reg/imm16) (dst reg))
  (behavior dst = s1 + s2)
  (format 000101 dst(25-21) s1(21-16) s2(15-0))
)
( opcode store
  (operands (s1 reg) (s2 imm16) (s3 reg))
  (behavior M[s1 + s2] = s3)
  (format 001101 s3(25-21) s1(21-16) s2(15-0))
)
```

Figure 2.7: Specification of the processor behavior using EXPRESSION ADL

The ADL also captures the mapping between the structure and the behavior (and vice versa). For example, the *add* and *sub* instructions are mapped to the *ALU* unit, the *load* and *store* instructions are mapped to the *MEM* unit, and so on.

2.3.2 Coprocessor Specification

The ADL specification of a programmable coprocessor is similar to the specification of the processor architecture described in Section 2.3.1. This section describes

[2] In this book we use the terms operation and instruction interchangeably.

how the ADL captures the structure and behavior of the coprocessor shown in Figure 2.4. To describe the structure of the coprocessor we specify each pipeline stage of the coprocessor along with the processor pipeline as shown in Figure 2.8(a). The coprocessor pipeline has three stages. The *EMIF_1* stage requests the DMA to transfer the data from the main memory to the coprocessor local memory. The *CoProc* stage performs the intended computation using the coprocessor local memory for accessing input operands. Results are stored back in the coprocessor memory. Finally, the *EMIF_2* requests the DMA to transfer the data from coprocessor memory to main memory. Figure 2.8(a) shows the description of the *CoProc* component. It supports four-cycle vector arithmetic operations.

```
# Components specification
......
( CPunit CoProc
  (capacity 1) (timing (vectAdd 4) (vectMul 4))
  (opcodes (vectAdd vectMul ...)) ...
)
# Pipeline and data-transfer paths
(pipeline Fetch Decode Execute MEM WriteBack)
(Execute (parallel ALU Coprocessor))
(Coprocessor (pipeline EMIF_1 CoProc EMIF_2))
(dtpaths (EMIF_1 DMA) (EMIF_2 DMA) ...)
......
```

(a) Structure

```
# Behavior: description of instruction-set
( opgroup cpOps
  (vectAdd, vectMul, ...)
)
......
( opcode vectMul
  (operands (s1 mem) (s2 mem) (dst mem) (length imm) )
  (behavior dst = s1 * s2)
)
```

(b) Behavior

Figure 2.8: Coprocessor specification using EXPRESSION ADL

The behavior of the coprocessor is captured in terms of the operations it supports. For example, Figure 2.8(b) shows the description of a *vectMul* operation. Unlike normal instructions whose source and destination operands are of type register (except load/store), here source and destination operands are of type memory. The *s1* and *s2* fields refer to the starting addresses of two source operands for the multiplication. Similarly *dst* refers to the starting address of the destination operand. The *length* field refers to the vector length of the operation that has immediate data type.

2.3.3 Memory Subsystem Specification

In order to explicitly describe the memory architecture in EXPRESSION, we need to capture both structure and behavior of the memory subsystem. The memory structure refers to the organization of the memory subsystem containing memory modules and the connectivity among them. The behavior refers to the memory subsystem instruction-set.

```
# Storage section
( DCache L1Data
  (wordsize 64) (linesize 8) (associativity 2)
  (num_lines 1024) (replacement LRU) (latency 1) ...
)
( ICache L1Inst (latency 1) ...)
( DCache L2 (latency 5) ...)
( DRAM MainMemory (latency 50) ...)
# Pipeline and data-transfer paths
(dtpaths (L1Data L2) (L1Inst L2) (L2 MainMemory) ...)
```

(a) Structure

```
# Behavior: description of instruction-set
( opcode load_miss
  (operands (s1 L2) (dst L1Data))
  (behavior dst = s1)
)
......
```

(b) Behavior

Figure 2.9: Memory subsystem specification using EXPRESSION ADL

The memory subsystem structure is represented as a netlist of memory components connected through ports and connections. The memory components are described and attributed with their characteristics (such as cache line size, replacement policy, and write policy). For example, Figure 2.9(a) shows the structure of the memory subsystem shown in Figure 2.4. The specification of the memory structure also includes the description of the memory pipeline and data-transfer paths. The memory subsystem instruction-set represents the possible operations that can occur in the memory subsystem, such as data transfers between different memory modules or to the processor (e.g., load and store) or explicit cache control instructions (e.g., prefetch, replace and refill). For example, Figure 2.9(b) shows an internal memory data transfer operation during a load miss. The *load_miss* operation represents data refill from L2 cache in the event of a L1 data miss.

2.4 Chapter Summary

This chapter surveyed existing ADLs in terms of their capabilities in capturing programmable architectures. Structural ADLs enable flexible and precise microarchitecture descriptions. The same description can be used for hardware synthesis, test generation, simulation and compilation. However, it is difficult to extract instruction-set information for retargetable compilation. Behavioral ADLs simplify the instruction-set description by exploiting the common components between operations. However, the lack of a detailed pipeline and timing information prevents the use of these languages as an extensible architecture model. Mixed ADLs capture both the structure and the behavior of the architecture.

The second part of this chapter described the use of the EXPRESSION ADL in our framework to specify programmable architectures. We described how to capture processor, coprocessor, and memory architectures using the ADL. Appendix C describes how to specify interrupts and exceptions in an ADL. Chapter 3 will present techniques to validate the ADL specification of the architecture.

3

VALIDATION OF SPECIFICATION

One of the most important requirements in a top-down validation methodology is to ensure that the specification (reference model) is golden. This chapter presents techniques to validate the static and dynamic behaviors of the architecture specified in an ADL. It is necessary to validate the ADL specification to ensure the correctness of both the architecture specified and the generated executable models including software toolkit and hardware implementation. The benefits of validation are two-fold. First, the process of any specification is error-prone and thus verification techniques can be used to check for correctness and consistency of the specification. Second, changes made to the processor during exploration may result in incorrect execution of the system and verification techniques can be used to ensure correctness of the modified architecture.

One of the major challenges in validating the ADL specification is to verify the pipeline behavior in the presence of hazards and multiple exceptions. There are many important properties that need to be verified to validate the pipeline behavior. For example, it is necessary to verify that each operation in the instruction-set can execute correctly in the processor pipeline. It is also necessary to ensure that execution of each operation is completed in a finite amount of time. Similarly, it is important to verify the execution style of the architecture. These properties are by no means complete to prove the correctness of the specification. Additional properties can easily be added and integrated into our validation framework.

The chapter is organized as follows. Section 3.1 describes the validation techniques to ensure that the static behavior of the pipeline is well-formed by analyzing the structural aspects of the specification using a graph based model. Section 3.2 presents the techniques to verify the dynamic behavior by analyzing the instruction flow in the pipeline using a FSM based model. Section 3.3 presents related work on validation of design specification. Finally, Section 3.4 summarizes the chapter.

3.1 Validation of Static Behavior

This section presents an automatic validation framework driven by an ADL. The first step (and only manual step) in the flow is to specify the architecture using an ADL such as EXPRESSION. A novel feature of this approach is the ability to model the pipeline structure and behavior of the processor, co-processor, and memory subsystem using a graph-based model. Based on this model, we present algorithms to ensure that the static behavior of the pipeline is well-formed by analyzing the structural aspects of the specification. Figure 3.1 shows the flow for validating static behaviors. The designer describes the programmable architecture in an ADL. The graph model of the architecture is generated from this ADL description. Several properties are applied to ensure that the architecture is well formed.

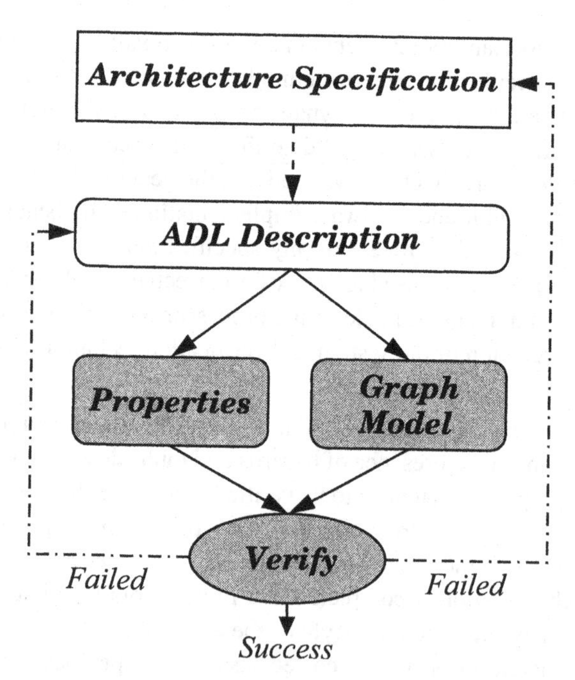

Figure 3.1: Validation of pipeline specifications

This section describes three important steps in this methodology. First, it presents a graph-based modeling of processor, memory, and co-processor pipelines. Second, it describes several properties that must be satisfied for valid pipeline specification. Finally, it illustrates validation of pipeline specifications for several realistic architectures.

3.1.1 Graph-based Modeling of Pipelines

We present a graph-based modeling of architecture pipelines that captures both the structure and the behavior. The graph model presented here can be derived from a pipeline specification of the architecture described in an ADL e.g., EXPRESSION [5]. This graph model can capture processor, memory, and co-processor pipelines for a wide variety of architectures including RISC, DSP, VLIW, and superscalar architectures. In this section, we briefly describe how the graph model captures the structure and behavior of the processor using the information available in the architecture manual.

Structure

The structure of an architecture pipeline is based on a block diagram view available in the processor manual, and is modeled as a graph $G_S = (V_S, E_S)$, where V_S denotes a set of components and E_S consists of a set of edges. V_S consists of two types of components: V_{unit} and $V_{storage}$. V_{unit} is a set of *functional units* (e.g., ALU), and $V_{storage}$ is a set of *storages* (e.g., register files). E_S consists of two types of edges. $E_{data\_transfer}$ is a set of *data-transfer edges*, and $E_{pipeline}$ is a set of *pipeline edges*. An edge (pipeline or data-transfer) indicates connectivity between two components. A data-transfer edge transfers data between units and storages. A pipeline edge transfers instruction (operation) between two units.

$$\begin{aligned} V_S &= V_{unit} \cup V_{storage} \\ E_S &= E_{data\_transfer} \cup E_{pipeline} \\ E_{data\_transfer} &\subseteq \{V_{unit},\, V_{storage}\} \times \{V_{unit},\, V_{storage}\} \\ E_{pipeline} &\subseteq V_{unit} \times V_{unit} \end{aligned}$$

For illustration, we use a simple multi-issue architecture consisting of a processor, a co-processor and a memory subsystem. Figure 3.2 shows the graph-based model of this architecture that can issue up to three operations (an ALU operation, a floating-point addition operation, and a coprocessor operation) per cycle. Figure 3.2 is obtained from Figure 2.4 by adding a four-stage floating point adder (*FADD*) and a feedback path from the *FADD* pipeline to the *ALU* pipeline. In the figure, oval boxes denote units, dotted ovals are storages, bold edges are pipeline edges, and dotted edges are data-transfer edges. A path from a root node (e.g., Fetch) to a leaf node (e.g, WriteBack) consisting of units and pipeline edges is called a *pipeline path*. For example, one of the pipeline paths is {*Fetch, Decode, ALU, MEM, WriteBack*}. A path from a unit to main memory or register file consisting of storages and data-transfer edges is called a *data-transfer path*. For example, {*MEM, L1Data, L2, MainMemory*} is a data-transfer path.

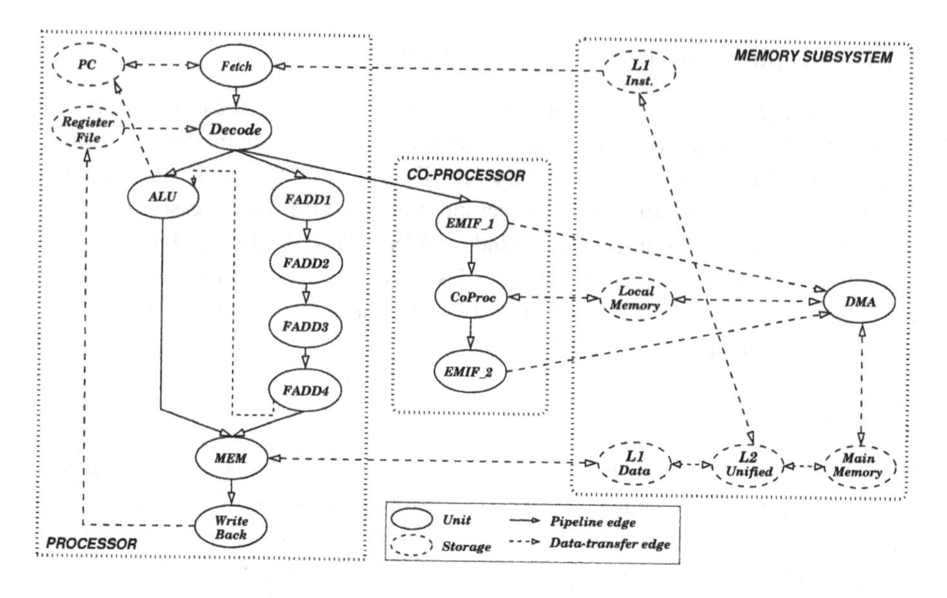

Figure 3.2: An example architecture

Behavior

The behavior of the architecture is typically captured by the instruction-set architecture (ISA) description in the processor manual. It consists of a set of operations that can be executed on the architecture. Each operation in turn consists of a set of fields (e.g. opcode, arguments) that specify, at an abstract level, the execution semantics of the operation. We model the behavior as a graph $G_B = (V_B, E_B)$, where V_B is a set of nodes, and E_B is a set of edges. The nodes represent the fields of each operation, while the edges represent orderings between the fields. The behavior graph G_B is a set of disjointed sub-graphs, and each sub-graph is called an *operation graph* (or simply an operation). Figure 3.3 shows a portion of the behavior (consisting of two operation graphs) for the example processor shown in Figure 3.2.

$$
\begin{aligned}
V_B &= V_{opcode} \cup V_{argument} \\
E_B &= E_{operation} \cup E_{execution} \\
E_{operation} &\subseteq V_{opcode} \times V_{argument} \ \cup \ V_{argument} \times V_{argument} \\
E_{execution} &\subseteq V_{argument} \times V_{argument} \ \cup \ V_{argument} \times V_{opcode}
\end{aligned}
$$

Nodes are of two types. V_{opcode} is a set of opcode nodes that represent the opcode (i.e. mnemonic), and $V_{argument}$ is a set of argument nodes that represent

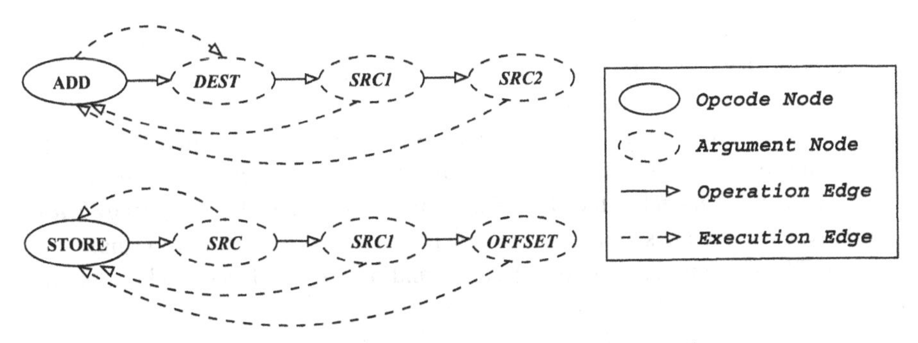

Figure 3.3: A fragment of the behavior graph

argument fields (i.e., source and destination arguments). In Figure 3.3, the ADD and STORE nodes are opcode nodes, while the others are argument nodes. Edges are also of two types. $E_{operation}$ is a set of operation edges that link the fields of the operation and also specify the syntactical ordering between them. On the other hand, $E_{execution}$ is a set of execution edges that specify the execution ordering between the fields. In Figure 3.3, the solid edges represent operation edges while the dotted edges represent execution edges. For the ADD operation, the operation edges specify that the syntactical ordering is opcode (ADD) followed by DEST, SRC1 and SRC2 arguments (in that order), and the execution edges specify that the SRC1 and SRC2 arguments are executed (i.e., read) before the ADD operation is performed. Finally, the DEST argument is written.

Mapping between Structure and Behavior

The mapping between the structure and the behavior is captured explicitly in the ADL. This information is available in the architecture manual as mapping between the instruction-set and the functional units. It other words, the manual describes what operations are supported by which functional units in the architecture. We define a set of mapping functions that map nodes in the structure to nodes in the behavior (and vice-versa). The *unit-to-opcode (opcode-to-unit)* mapping is a bi-directional function that maps unit nodes in the structure to opcode nodes in the behavior. The *unit-to-opcode* mappings for the architecture in Figure 3.2 include mappings from *Fetch* unit to opcodes {*ADD, FADD*}, *ALU* unit to opcode *ADD*, *FADD1* unit to opcode *FADD* etc. The *argument-to-storage (storage-to-argument)* mapping is a bi-directional function that maps argument nodes in the behavior to storage nodes in the structure. For example, the *argument-storage* mappings for the *ADD* operation are mappings from {*DEST, SRC1, SRC2*} to *RegisterFile*.

Each functional unit (with read or write ports) supports certain data-transfer operations. These operations can be derived from the above mapping functions. For example, the *Decode* unit of Figure 3.2 supports register read (*regRead*) for ADD and LD opcodes; the *MEM* unit supports data read (*dataRead*) and data write (*dataWrite*) from L1 data cache; the *Fetch* unit supports instruction read (*instRead*) from L1 instruction cache; the *WriteBack* unit supports register write (*regWrite*). Similarly, each storage supports certain data-transfer operations. For example, the *RegisterFile* of Figure 3.2 supports *regRead* and *regWrite*; L1 data cache supports *dataRead* and *dataWrite*, and so on.

3.1.2 Validation of Pipeline Specifications

Based on the graph model presented in the previous section, the ADL specification of the architecture pipelines can be validated. In this section, we describe some of the properties used in our framework for validating pipelined architecture specifications.

Connectedness Property

The connectedness property ensures that each component is connected to other component(s). As pipeline and data-transfer paths are connected regions of the architecture, this property holds if each component belongs to at least one pipeline or data-transfer path.

$$\forall v_{comp} \in V_S, (\exists G_{PP} \in \mathbf{G_{PP}},\ s.t.\ v_{comp} \in G_{PP}) \vee (\exists G_{DP} \in \mathbf{G_{DP}},\ s.t.\ v_{comp} \in G_{DP})$$

where $\mathbf{G_{PP}}$ is a set of pipeline paths and $\mathbf{G_{DP}}$ is a set of data-transfer paths.

Algorithm 1 presents the pseudo-code for verifying the connectedness property. The algorithm requires the graph model G of the architecture as input. It also requires all the component lists as input. The first step is to unmark the entries in all the input lists. Each input list contains all the respective components in the graph. For example, the *ListOfUnits* contains all the units in the graph G. Next, the graph is traversed in breadth-first manner and the visited components are marked. For example, when a unit u is visited during traversal, it is marked in *ListOfUnits*. Finally, the algorithm returns true if all the entries are marked in all the input lists. It returns false if there are any unmarked entries in any of the input lists, and it reports them. Each node of the graph is visited only once. The time and space complexity of the algorithm is $O(n)$, where n is the number of nodes in the graph G. Each node of the graph can be either unit or storage.

```
Algorithm 1: Verify Connectedness
Inputs: i. Graph model of the architecture G
        ii. ListOfUnits: list of units in the graph G
        iii. ListOfStorages: list of storages in the graph G
Outputs: i. True, if the graph model satisfies this property else false.
         ii. In case of failure, report the disconnected components.
Begin
    Unmark all the entries in all the input lists.
    InsertQ(root, Q) /* Put root node of G in queue Q */
    while Q is not empty
        Node n = DeleteQ(Q) /* Remove the front element of Q */
        Mark n as visited in G
        case type of node n
            unit: Mark n in ListOfUnits
            storage: Mark n in ListOfStorages
        endcase
        for each successor node s of n
            if s is not visited InsertQ(s, Q)
        endfor
    endwhile
    Return true if all the entries are marked in all of the input lists;
        false otherwise, and report the unmarked components.
End
```

False Pipeline and Data-transfer Paths

According to the definition of pipeline paths, there may exist pipeline paths that are never activated by any operation. Such pipeline paths are said to be *false*. For example, consider the architecture shown in Figure 3.4 that executes two operations: ALU-shift (*alus*) and multiply-accumulate (*mac*). This processor has unit-to-opcode mappings between ALU unit and opcode *alus*, between SFT unit and *alus*, between MUL unit and *mac*, and between ACC unit and *mac*. Also, there are unit-to-opcode mappings between each of {*IFD, RD1, RD2, WB*} and *alus*, and each of {*IFD, RD1, RD2, WB*} and *mac*. This processor has four pipeline paths: {*IFD, RD1, ALU, RD2, SFT, WB*}, {*IFD, RD1, MUL, RD2, ACC, WB*}, {*IFD, RD1, ALU, RD2, ACC, WB*}, and {*IFD, RD1, MUL, RD2, SFT, WB*}. However, the last two pipeline paths cannot be activated by any operation. Therefore, they are false pipeline paths. Since these false pipeline paths may become false paths depending on the detailed structure of RD2, they should be detected at a higher level of abstraction to ease the later design phases.

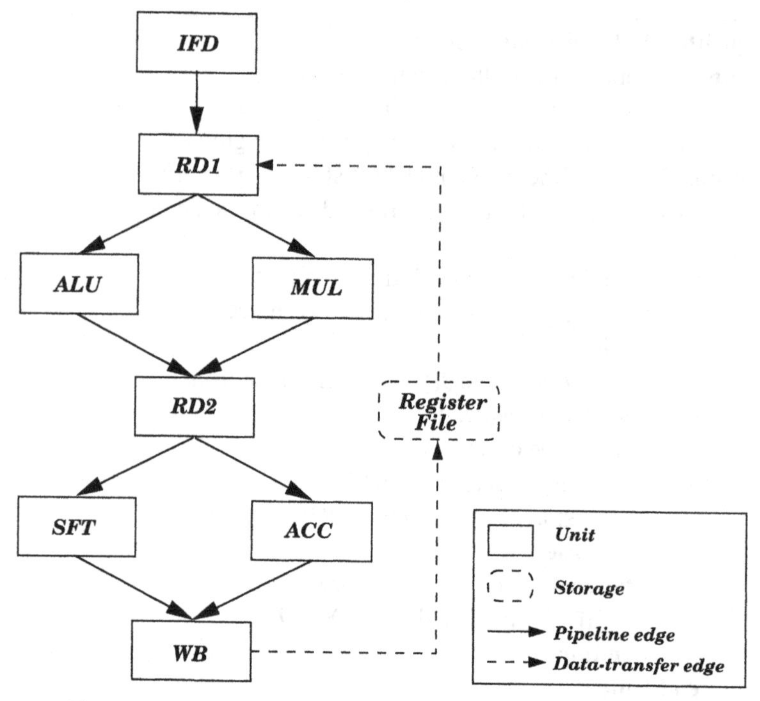

Figure 3.4: An example processor with false pipeline paths

From the view point of SOC architecture exploration, we can view the false pipeline paths as an indication of potential behaviors that are not explicitly defined in the ADL description. These false pipeline paths can be used to perform valid computations. This opens up avenues for further exploration experiments for cost, power, and performance by adding new instructions to activate the false pipeline paths. Formally, a pipeline path $G_{PP}(V_{PP}, E_{PP})$ is false if the intersection of opcodes supported by the units in the pipeline path is empty.

$$\bigcap_{v_{unit} \in V_{PP}} f_{unit-opcode}(v_{unit}) = \phi \tag{3.1}$$

Similarly, there may exist data-transfer paths in the specification that are never activated by any operation. Such data-transfer paths are said to be *false*. For example, consider the architecture shown in Figure 3.5 that has seven possible data-transfer operations: integer-register-read (IregRd), float-register-read (FregRd), integer-register-write (IregWr), float-register-write (FregWr), load-data-from-memory (ldData), load-instruction-from-memory (ldInst), and store-data-in-memory (stData). The *Decode (ID)* unit has mappings for *IregRd* and *FregRd*. There are mappings between each of {*WB1, WB2*} and {*IregWr, FregWr*}, each of {*IF, L1I, ISB*} and *ldInst*, each of {*LDST, L1D, DSB*} and {*ldData, stData*}, and each of {*L2, DRAM*} and {*ldData, stData, ldInst*}. This processor has ten data-transfer paths: {*IRF, ID*}, {*FRF, ID*}, {*WB1, IRF*}, {*WB1, FRF*}, {*WB2, IRF*}, {*WB2, FRF*}, {*IF, L1I, L2, ISB, DRAM*}, {*LDST, L1D, L2, DSB, DRAM*}, {*IF, L1I, L2, DSB, DRAM*}, {*LDST, L1D, L2, ISB, DRAM*} . However, the last two data-transfer paths cannot be activated by any operation. Therefore, they are false data-transfer paths. If *ALU1* supports only floating-point operations, the fourth path ({*WB1, IRF*}) becomes a false data-transfer path.

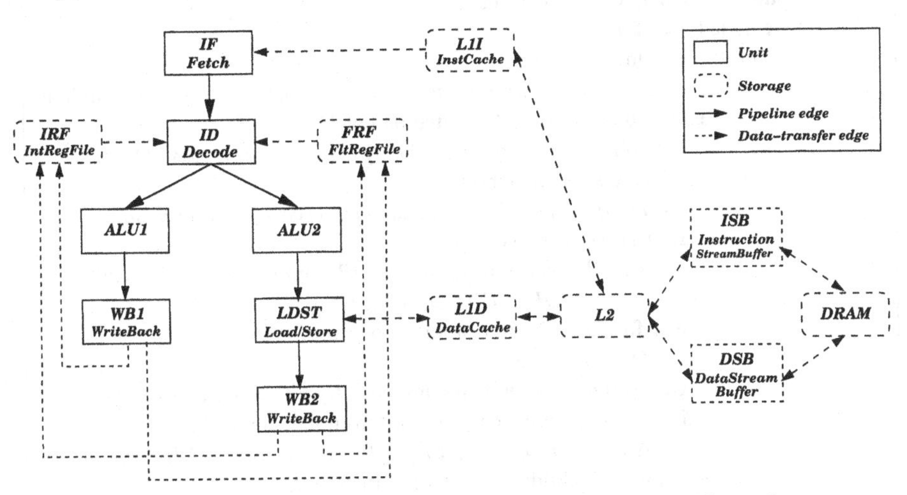

Figure 3.5: An example processor with false data-transfer paths

Formally, a data-transfer path $G_{DP}(V_{DP}, E_{DP})$ is false if the intersection of data-transfer operations supported by the units and storages ($f_{node-operation}$) in the data-transfer path is empty.

$$\bigcap_{v_{node} \in V_{DP}} f_{node-operation}(v_{node}) = \phi \tag{3.2}$$

Algorithm 2 presents the pseudo-code for verifying false pipeline and data-transfer paths. The algorithm requires the graph model G as input. It traverses the graph in depth-first manner along each pipeline and data-transfer path. Each unit u has a list of supported opcodes $SuppOpList_u$. Each node n (unit or storage) also maintains four temporary lists: $OutOpList_n$, $OutDTopList_n$, $InOpList_n$, and $InDTopList_n$. The $OutOpList_n$ is the list of opcodes produced by unit n and sent to its children units. The $OutDTopList_n$ is the list of data-transfer operations produced by node (unit or storage) n and sent to its children storages. The $InOpList_n$ is the list

```
Algorithm 2: Verify False Pipeline and Data-transfer Paths
Input: Graph model of the architecture G.
Outputs: i. True, if the graph model satisfies this property else false.
         ii. In case of failure, report the list of false pipeline and data-transfer paths.
{
  Push (root, S); FalsePPpathList = {}; FalseDPpathList = {};
  while S is not empty
      Node n = Pop(S); Mark n as visited.
      case node type of n
          unit: if n is the root node
                    OutOpList_n = SuppOpList_n // Send supported opcodes to children
                else /* p is the recently visited parent */
                    InOpList_n = OutOpList_p; OutOpList_n = SuppOpList_n ∩ InOpList_n
                if n has read or write ports
                    OutDTopList_n = ComputeDataTransferOps(OutOpList_n)
                    if OutDTopList_n is empty
                        for all the data-transfer paths fDP from n to any leaf nodes
                            Insert fDP in FalseDPpathList.
                    else for each children storage node st of n, Push(st, S)
                if OutOpList_n is empty
                    Get path pp from n by tracing recently visited parents till root
                    for all the pipeline paths ppEnd from n to any leaf nodes
                        Append ppEnd to pp to get fPP; Insert fPP in FalsePPpathList.
                else for each children unit u of n, Push(u, S)
          storage: InDTopList_n = OutDTopList_p
                OutDTopList_n = SuppDTopList_n ∩ InDTopList_n
                if OutDTopList_n is empty
                    Get path dp from n by tracing recently visited parents till any unit
                    for all the data-transfer paths dpEnd from n to any leaf nodes
                        Append dpEnd to dp to generate false data-transfer path fDP.
                        Insert fDP in FalseDPpathList.
                    endfor
                else for each children storage node st of n, Push(st, S)
      endcase
  endwhile
  if FalsePPpathList and FalseDPpathList are empty return true;
  else return false and report FalsePPpathList and FalseDPpathList.
}
```

that is used by unit n to copy the $OutOpList_p$, the output list of the recently visited parent p. Similarly, the $InDTopList_n$ is the list that is used by storage n to copy the $OutDTopList_p$, the output list of the recently visited parent p. Each unit n performs intersection of $InOpList_n$ and $SuppOpList_n$ and send the result $OutOpList_n$ to its children units. If $OutOpList_n$ is empty, all the pipeline paths that use the path from n to root (via recently visited parents) are false pipeline paths. A unit with read or write ports computes data transfer operations using the method described in Section 3.1.1. A storage computes $OutDTopList_n$ by performing intersection of $SuppDTopList_n$ and the input list $InDTopList_n$. If $OutDTopList_n$ is empty, all the data-transfer paths that use the path from storage n to any unit via recently visited parents are false data-transfer paths. The algorithm returns true if there are no false pipeline or data-transfer paths. It returns false if there are any false pipeline or data-transfer paths, and reports them.

If there are n nodes, x pipeline and data-transfer paths, and p operations (opcodes) supported by the processor, the time complexity of the algorithm is $O(x \times n \times (x + plogp))$ and space complexity is $O(n \times p)$. The supported opcode list in each node is a sorted list.

Completeness Property

The completeness property confirms that all operations must be executable. An operation op is executable if there exists a pipeline path $G_{PP}(V_{PP}, E_{PP})$ on which op is executable i.e., if both Condition 3.1 and 3.2 hold.

Condition 3.1: All units in V_{PP} support the operation op. More formally, the following condition holds where v_{opcode} is the opcode of the operation op.

$$\forall v_{unit} \in V_{PP}, v_{opcode} \in f_{unit-opcode}(v_{unit}). \quad (3.3)$$

Condition 3.2: There are no conflicting partial orderings of operation arguments and unit ports. Let V be a set of argument nodes of operation op. There are no conflicting partial orderings of operation arguments and unit ports if, for any two nodes $v_1, v_2 \in V$ such that $(v_1, v_2) \in E_{execution}$, all the following conditions hold:

- There exists a data-transfer path from a storage $f_{arg-storage}(v_1)$ to a unit v_{u1} in V_{PP} through a port $f_{arg-port}(v_1)$.
- There exists a data-transfer path from a unit v_{u2} in V_{PP} to a storage $f_{arg-storage}(v_2)$ through a port $f_{arg-port}(v_2)$.
- v_{u1} and v_{u2} are the same unit or there is a path consisting of pipeline edges from v_{u1} to v_{u2}.

```
Algorithm 3: Verify Completeness
Inputs: i. Graph model G of the architecture.
        ii. The list of operations OpList supported by the architecture.
Outputs: i. True, if the graph model satisfies this property else false.
         ii. In case of failure, report the list of operations that are not executable.
{
 for each operation op supported by the architecture /* op ∈ OpList */
    opSrcList = list of sources in the operation op.
    opDestList = list of destinations in the operation op.
    Push(root, S) /* Put root node of G in stack S */
    while S is not empty
          Node n = Pop(S); Mark n as visited in G.
          if op ∈ SuppOpList_n /* op is supported by unit n */
            for each port p of n
               if p is a read or read-write port
                 for each unmarked source src in opSrcList
                     if src can be read via p, mark src in opSrcList with (p, n)
               if p is a write or read-write port
                 for each unmarked destination dest in opDestList
                     if dest can be written via p
                        mark dest in opDestList with (p, n)
            endfor
            if unit n is a leaf node
              if ( ( all sources in opSrcList are marked ) and
                  ( all nodes r that read the sources are in expected order) and
                  ( all destinations in opDestList are marked ) and
                  ( all nodes w that write the destinations are in expected order) and
                  ( all nodes r & w are in same pipeline path and r appears before w ))
                Mark op in OpList /* this path supports op */
                break /* one pipeline path is sufficient, exit while loop */
              endif
            else for each children unit u of n, Push(u, S)
    endwhile
 endfor
 Return true if all the entries in OpList are marked;
        false otherwise, and report the unmarked entries in OpList.
}
```

For example, let us consider the *ADD* operation (shown in Figure 3.3) for the processor described in Figure 3.2. To satisfy Condition 3.1, each of {*Fetch, Decode, ALU, MEM, WriteBack*} must have mappings to the *ADD* opcode. On the other hand, Condition 3.2 is satisfied because the structure has data-transfer paths from *RegisterFile* to *Decode* and from *WriteBack* to *RegisterFile*, and there is a pipeline path from *Decode* to *WriteBack*.

Algorithm 3 presents the pseudo-code for verifying the completeness property. The algorithm requires the graph model (*G*) and the list of operations supported by the architecture (*OpList*) as inputs. It traverses the graph in depth-first manner for each operation *op* and identifies a pipeline path *pp* that supports *op*. All the units *n* in the pipeline path should have *op* in their supported opcode list $SuppOpList_n$. The pipeline path *pp* must have units that can read the source operands of *op* and write the destination operands of *op* in correct order. If all the conditions are met, *op* is executable in pipeline path *pp* and *op* is marked in *OpList*. The algorithm returns true if all the entries in *OpList* are marked. It returns false if there are unmarked entries and reports them.

If there are *n* nodes, *x* pipeline and data-transfer paths in the graph and the number of opcodes supported by the architecture is *p*, the time complexity of the algorithm is $O(x \times n \times p \times logp)$ and space complexity is $O(n \times p)$. The opcode list in each unit is a sorted list.

Finiteness Property

The finiteness property guarantees the termination of any operation executed through the pipeline. The termination is guaranteed if all pipeline and data-transfer paths except false pipeline and data-transfer paths have finite length and all nodes on the pipeline or data-transfer paths have finite timing. The length of a pipeline or data-transfer path is defined as the number of stages required to reach the final (leaf) nodes from the root node of the graph. Formally,

$$\exists K,\ s.t.\ \forall path \in (\mathbf{G_{PP}}, \mathbf{G_{DP}}), num\_stages(path) < K \tag{3.4}$$

Here, *num_stages* is a function that, given a pipeline or data-transfer path, returns the number of stages (i.e. clock cycles) required to execute it. In the presence of cycles in the pipeline path, this function cannot be determined from the structural graph model alone. However, if there are no cycles in the pipeline paths, the termination property is satisfied if the number of nodes in V_S is finite, and each multi-cycle component has finite timing.

Algorithm 4 presents the pseudo-code for verifying finiteness property. The algorithm requires the graph model *G* and the list of operations supported by the architecture (*OpList*) as inputs. It traverses the graph in depth-first manner for each operation *op* and identifies all the pipeline paths *op-pp* that support *op*. For each operation it marks different pipeline paths *op-pp* with a different color. A cycle is detected if the same colored node is visited more than once during traversal.

```
Algorithm 4: Verify Finiteness
Inputs: i. Graph model G of the architecture.
        ii. The list of operations OpList supported by the architecture.
Outputs: i. True, if the graph model satisfies this property else false.
         ii. In case of failure, report the list of paths that violates this property.
{
  PathList = {};
  for each operation op supported by the architecture
     PathLength = 0; ColorCode = 0
     Push(< root, PathLength >, S); Unmark all the nodes in graph G
     while S is not empty
          < n, PathLength > = Pop(S)
          if op ∈ SuppOpList_n /* op is supported by unit n */
            PathLength = PathLength + 1; timing = GetExecutionTime(op, n);
            if ( ( n is already marked with ColorCode) or
                 ( timing is greater than MaxExecutionTime) or
                 ( PathLength is greater than MaxPathLength ) )
               Insert < op, path > pair in PathList; break; /* exit while loop */
            else
               Mark n with ColorCode
               if unit n is a leaf node, ColorCode = ColorCode + 1;
               else
                 for each children node child of n
                     Push(< child, PathLength >, S);
               endif
            endif
          else ColorCode = ColorCode + 1;
     endwhile
  endfor
  Return true if PathList is empty
         false otherwise, and report PathList.
}
```

The pipeline path *op-pp* with cycle will be stored in *PathList*. This property is also violated when there are paths that are longer than *MaxPathLength* or when the execution time needed by *op* in any node in that path is greater than *MaxExecutionTime*. The algorithm returns true if *PathList* is empty. It returns false if there are entries in *PathList* and reports them.

Our finiteness algorithm assumes that there are no cycles in the pipeline. If the cycles are allowed in the pipeline due to the reuse of the resources, our algorithm needs to be modified. Let us assume that a resource is reused by an operation *op* for n_{op} times. We can modify the algorithm to check for *"already marked with ColorCode for* n_{op} *times"* instead of checking *"already marked with ColorCode"* for the operation *op*. If there are *n* nodes, *x* pipeline and data-transfer paths in the graph and the number of opcodes supported by the architecture is *p*, the time complexity of this algorithm is $O(x \times n \times p \times logp)$ and space complexity is $O(n \times p)$. The opcode list in each unit is a sorted list.

Architecture-specific Properties

The architecture must be well-formed based on the original intent of the architecture model. Here we mention some of the architecture specific properties we verify in our framework.

- ❒ The number of operations processed per cycle by a unit can not be smaller than the total number of operations sent by its parents unless the unit has a reservation station. This event (fewer output instructions than the input instructions) is not an error if that specific unit kills certain operations based on certain conditions e.g., killing no operation (NOP).

- ❒ The instruction template should match the available pipeline bandwidth. However, having instruction template size different than pipeline bandwidth does not always imply an error because a machine with *n* operations in an instruction and *m* ($> n$) parallel pipeline paths may have many multicycle units. Similarly, the architecture may have *m* ($< n$) parallel pipeline paths if it has a reservation station and the instruction fetch timing is large.

- ❒ There must be a path from load/store unit to main memory via storage components to ensure that every memory operation is complete.

- ❒ The address space used by the processor must be equal to the union of the address spaces covered by memory subsystem (SRAM, cache hierarchies etc.) to ensure that all the memory accesses can complete.

Algorithm 5 shows how we apply these properties in our framework. We first verify finiteness property before applying any other properties in our framework. If there are paths with infinite length and timing, the finiteness algorithm will display the paths and exit. Next, we apply the connectedness property followed by the false pipeline and data-transfer path property. The remaining properties can be applied in any order. The worst case time complexity of Algorithm 5 is $O(x \times n \times (x + plogp))$ and space complexity is $O(n \times p)$, where the architecture graph has n nodes, x pipeline and data-transfer paths, and the number of operations supported by the processor is p. Typically, the numeric values of these variables are not large: both n and x are less than 100, and p is less than 1000. As a result, it requires less than a second to verify an architecture specification as demonstrated in the next section.

```
Algorithm 5: Verify Architecture Specification
Input: Graph model G of the architecture.
Output: True, if the graph model satisfies all the properties
          else false, and report the error.
Begin
      status = VerifyFiniteness (G, G.SupportedOpcodeList);
      if (status == false)
        Report the paths that violate this property;
        return false;
      endif
      status = VerifyConnectedness (G, G.ListOfUnits, ... );
      if (status == false)
        Report the components that are not connected;
        return false;
      endif
      status = VerifyFalsePipelineDataTransferPaths(G);
      if (status == false)
        Report the list of false pipeline and data-transfer paths;
        return false;
      endif
      status = VerifyCompleteness (G, G.SupportedOpcodeList);
      if (status == false)
        Report the list of operations that are not executable;
        return false;
      endif
      /* Apply other architecture specific properties */
      .........
      return true;
End
```

3.1.3 Experiments

In order to demonstrate the applicability and usefulness of our validation approach, we have described a wide range of architectures using the EXPRESSION ADL: MIPS R10K [50], TI C6x [131], PowerPC [48], and DLX [55] that are collectively representative of RISC, DSP, VLIW, and superscalar architectures. Our framework generates the graph model from the ADL specification. We have implemented each property as a function that operates on the graph model. Finally, we have applied these properties on the graph model to verify that the specified architecture is well-formed. Table 3.1 shows the specification validation time for different architectures on a 333 MHz Sun Ultra-II with 128M RAM. This includes the time to generate the graph model from the ADL specification and to apply all the properties on the graph model. The validation time depends on three aspects: number of properties applied, complexity of the structure and the number of operations supported by the architecture. Typically, the validation time is in the order of seconds.

Table 3.1: Specification validation time for different architectures

Architecture	DLX	TI C6x	PowerPC	MIPS R10K
Validation Time (sec)	0.1	0.2	0.3	0.5

In the remainder of this section, we describe our specification validation experiments. First, we describe the validation of the DLX specification in detail. Next, we summarize the incorrect specification errors captured by our framework during design space exploration of different architectures.

Validation of the DLX specification

Our framework generated the graph model from the ADL specification of the DLX architecture. Figure 3.6 shows the simplified graph model of the DLX architecture. Figure 3.6 is obtained by adding two execution paths (seven-stage multiplier and a multi-cycle divider) in the processor pipeline shown in Figure 3.2. The oval (unit) and rectangular (storage) boxes represent nodes. The solid (pipeline) and dotted (data-transfer) lines represent edges.

We applied all the properties (Algorithm 5) on the graph model. We encountered two kinds of errors viz., incomplete specification errors and incorrect specification errors. An example of an incomplete specification error we uncovered is that the opcode assignment is not done for the fifth stage of the multiplier pipeline. Similarly, an example of an incorrect specification error we found is that only load/store opcodes were mapped for the memory stage (*MEM*). Since all the opcodes pass through memory stage in DLX, it is necessary to map all the opcodes in memory stage as well.

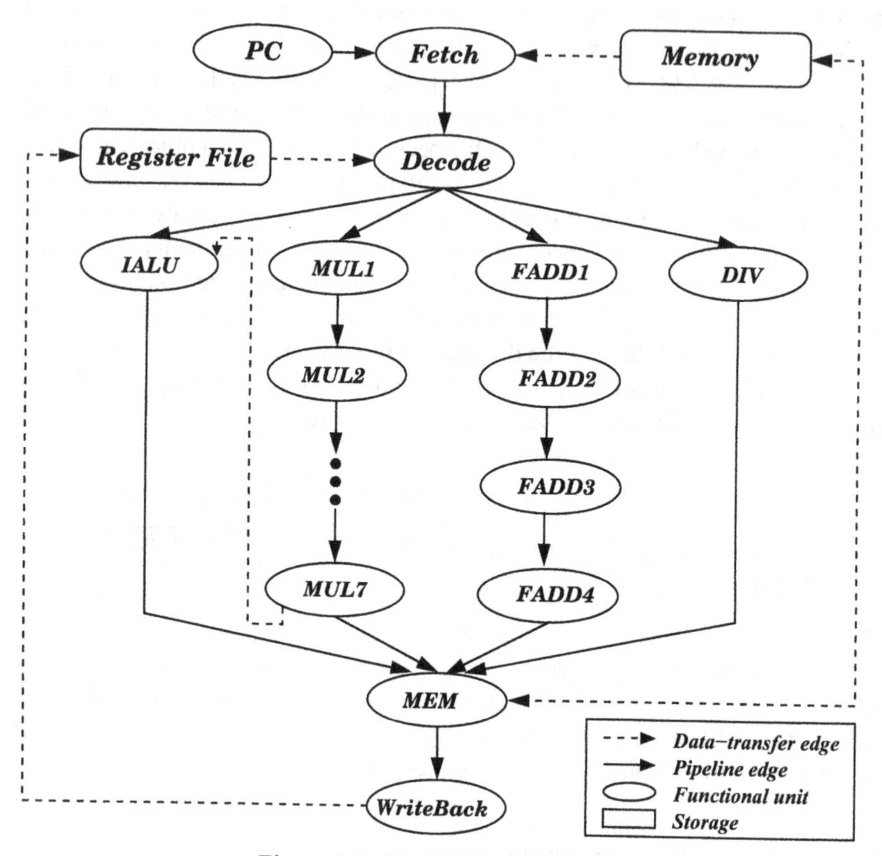

Figure 3.6: The DLX architecture

We used Algorithm 5 for specification validation. First, the finiteness property is applied on the graph model. It detects a violation for the division operation since the multi-cycle division unit (*DIV*) has an undefined latency value. Once the latency for the division operation is defined, the finiteness property is successful. Next, the connectedness property is applied. It detects that the sixth stage of the multiplier unit (*MUL6*) is not connected. Once it is connected properly (from *MUL5* to *MUL6*, and from *MUL6* to *MUL7*), the connectedness property is successful. Finally, the completeness property is violated for the multiply operation. This operation is not defined in the *MUL5* unit. As a result, the multiply operation cannot execute in the pipeline. Once this is fixed, the validation of the DLX specification is successful.

Violation of Properties during DSE

We have performed many modifications of several architecture specifications. These are typical changes made by a designer during exploration. It is important to note that although a designer may think that changes made during exploration may not introduce errors, it is often the case that several subtle errors may be introduced during changes to the architecture, which highlights the need for the validation of the modified specification before proceeding with other design steps. Here we briefly mention some of the errors that were captured using our approach during some typical architectural exploration scenarios.

- ☆ We have modified the MIPS R10K ADL description to add another load/store unit that supports only store operations. The false data-transfer path property is violated since there is a write connection from the load/store unit to the floating-point register file that will never be used.
- ☆ We have modified the PowerPC ADL description to have a separate L2 cache for instruction and data. Validation determined that there were no paths from L2 instruction cache to main memory. The connection between L2 instruction cache and unified L3 cache was missing.
- ☆ We have modified the C6x architecture's data memory by adding two SRAM modules to the existing cache hierarchy. The property validation fails due to the fact that the address ranges specified in the SRAMs and cache hierarchy are not disjoint, moreover union of these address ranges did not cover the physical address space specified by the processor description.
- ☆ In the R10K architecture we have decided to use a coprocessor local memory instead of integer register file for reading operands. We have removed the read connections that are used to access the integer register file and added local memory, DMA controller and connections to the main memory. The connectedness property is violated for two ports in the integer register file.

Table 3.2 summarizes the errors captured during design space exploration of architectures. Each column represents one architecture and each row represents one property. An entry in the table presents the number of violations of that property for the corresponding architecture[1]. The number (in parenthesis) below each architecture represents the number of design space explorations done for that architecture. Each class of problem is counted only once. For example, the DLX error

[1]The error numbers will change depending on the number of design space explorations and the type of modifications done each time.

mentioned above (where one of the unit has incorrect specification of the supported opcodes that led to false pipeline path for most of the opcodes) is counted only once instead of using the number of opcodes that violated the property.

Table 3.2: Summary of property violations during DSE

	DLX (2)	C6x (2)	R10K (3)	PowerPC (2)
Connectedness	0	1	2	1
False Pipeline/Data-transfer Path	5	3	4	2
Completeness	2	3	3	2
Architecture-specific	4	5	12	6
Finiteness	0	0	1	1

Our experiments have demonstrated the utility of our validation approach across a wide range of realistic architectures, and the ability to detect errors in the architecture specification, as well as errors generated through inconsistent modifications to an architecture during design space exploration.

3.2 Validation of Dynamic Behavior

This section presents novel techniques to verify the dynamic behavior of an architecture specified in an ADL by analyzing the instruction flow in the pipeline. Figure 3.7 shows our modified methodology for validation of static and dynamic behaviors. The FSM model is generated from the ADL specification. Based on this model, we propose a method for validating pipelined processor specifications using two properties: determinism and in-order execution. The remainder of this section is organized as follows. First, we describe a FSM-based modeling of processor pipelines. Next, we present the validation technique followed by a case study using the DLX architecture.

3.2.1 FSM-based Modeling of Processor Pipelines

In this section we describe how we derive the FSM model of the processor pipeline from the ADL specification. We first explain how we specify the information necessary for FSM modeling, then we present the FSM model of the processor pipelines using the information captured in the ADL.

A. Processor Pipeline Description in ADL

Figure 3.8 shows a fragment of a processor pipeline. The oval boxes represent units, rectangular boxes represent pipeline latches, and arrows represent pipeline edges. In this section we briefly describe how we specify pipeline flow conditions for stalling, normal flow, bubble insertion, exception and squashing in the ADL.

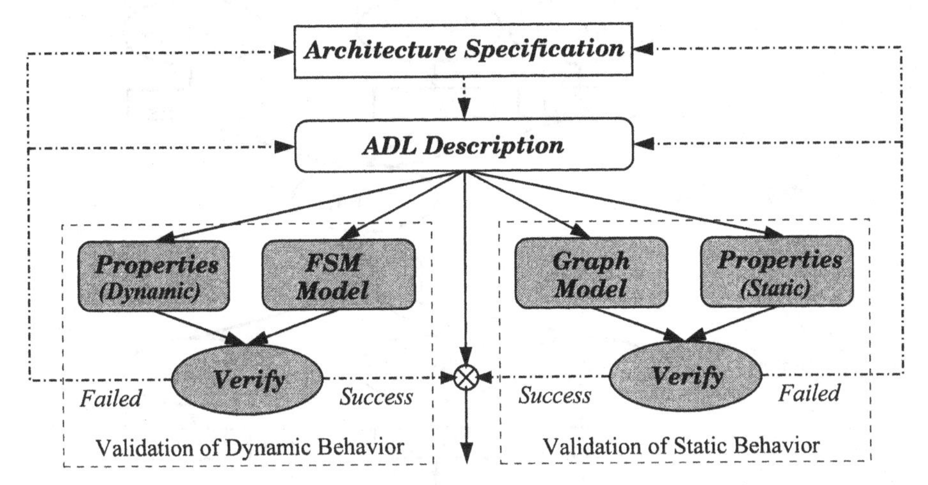

Figure 3.7: ADL driven validation of pipeline specifications

A unit is in *normal flow* (NF) if it can receive instruction from its parent unit and can send to its child unit. A unit can be *stalled* (ST) due to external signals or due to conditions arising inside the processor pipeline. For example, the external signal that can stall a fetch unit is cache miss; the internal conditions to stall the fetch unit can be due to decode stall, hazards, or exceptions. A unit performs *bubble insertion* (BI) when it does not receive any instruction from its parent (or busy computing in case of multicycle unit) and its child unit is not stalled. A unit can be in *exception* condition due to internal contribution or due to an exception. A unit is in bubble/nop *squashed* (SQ) stage when it has a nop instruction that gets removed or overwritten by an instruction of the parent unit.

For units with multiple children the flow conditions due to internal contribution may differ. For example, the unit $UNIT_{i-1,j}$ in Figure 3.8 with q children can be *stalled* when *any* one of its children is stalled, or when *some* of its children are stalled (designer identifies the specific ones), or when *all* of its children are stalled; or when *none* of its children are stalled. During specification, the designer selects from the set {*any, some, all, none*} the internal contribution along with any external signals to specify the stall condition for each unit. Similarly, the designer specifies the internal contribution for other flow conditions [95].

The PC unit can be *stalled* (ST) due to external signals such as cache miss or when the fetch unit is stalled. When a branch is taken the PC unit is said to be in *branch taken* (BT) state. The PC unit is in *sequential execution* (SE) mode when the fetch unit is in normal flow, there are no external interrupts, and the current instruction is not a branch instruction.

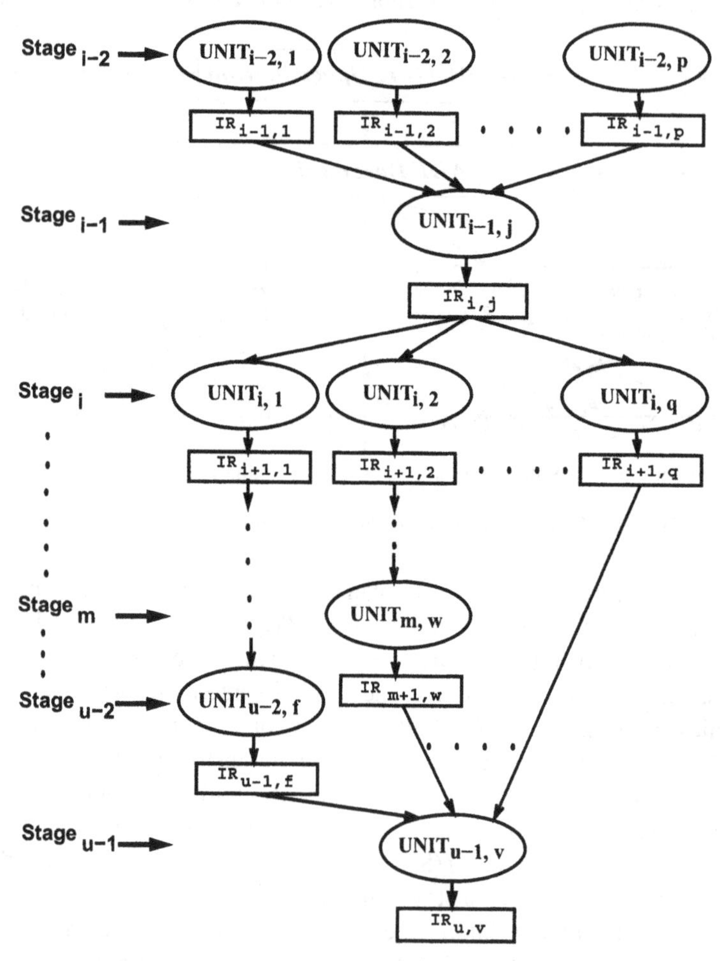

Figure 3.8: A fragment of a processor pipeline

B. FSM Model of Processor Pipelines

This section presents an FSM-based modeling of controllers in pipelined processors. Intuitively, the FSM captures the information of all the storage elements in

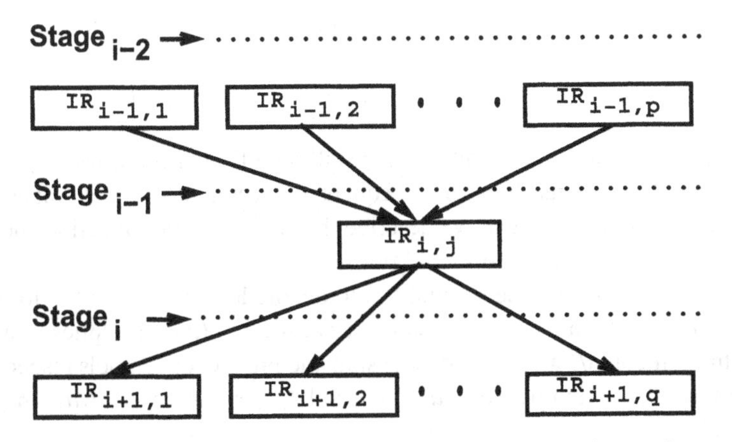

Figure 3.9: The processor pipeline with only instruction registers

the pipeline including program counter and pipeline latches. Let us assume that there are n such elements. Therefore, a state S_t^n in the FSM has the values of all the n elements at time t. The state transition (next-state) function returns the set of values of all the n elements at next time step (clock cycle). In other words the next state of S_t^n is S_{t+1}^n. The remainder of this section describes the FSM model in detail.

Figure 3.9 shows a fragment of the processor pipeline with only instruction registers[2] (IR). We assume a pipelined processor with in-order execution as the target for modeling and validation. The pipeline consists of n stages. Each stage can have more than one pipeline register (in case of fragmented pipelines). Each single-cycle pipeline register takes one cycle if there are no pipeline hazards. A multi-cycle pipeline register takes m cycles during normal execution (no hazards). Let $Stage_i$ denote the i-th stage where $0 \leq i \leq n-1$, and n_i is the number of pipeline registers between $Stage_{i-1}$ and $Stage_i$. Let $IR_{i,j}$ denote an instruction register between $Stage_{i-1}$ and $Stage_i$ ($1 \leq i \leq n$, $1 \leq j \leq n_i$). The first stage, i.e., $Stage_0$, fetches an instruction from the instruction memory pointed by the program counter PC, and stores the instruction into the first instruction register $IR_{1,j}$ ($1 \leq j \leq n_1$). Without loss of generality, let us assume that $IR_{i,j}$ has p parent units and q children units as shown in Figure 3.9. During execution, the instruction stored in $IR_{i,j}$ is executed at $Stage_i$ and then stored into the next instruction register $IR_{i+1,k}$ ($1 \leq k \leq q$).

In this section, we define a state of the n-stage pipeline as values of PC and $\sum_{i=1}^{n-1} n_i$ instruction registers. Let $PC(t)$ and $IR_{i,j}(t)$ denote the values of PC and

[2]We refer to these pipeline latches (registers) as instruction registers since they are used to transfer instructions from one pipeline stage to the next.

$IR_{i,j}$ at time t, respectively. Then, the state of the pipeline at time t is defined as

$$S(t) = < PC(t), IR_{1,1}(t), \cdots, IR_{i,j}(t), \cdots, IR_{n-1,n_{n-1}}(t) > \quad (3.5)$$

We first describe the flow conditions for stalling(ST), normal flow(NF), bubble insertion(BI), bubble squashing (SQ), sequential execution(SE), and branch taken (BT) in the FSM model. Next, we describe the state transition functions possible in the FSM model using the flow conditions.

In this section we use the symbol '$\vee$' to denote logical *or*, and '$\wedge$' to denote logical *and*. For example, $(a \vee b)$ implies (a or b), and $(a \wedge b)$ implies (a and b). We use the symbols $\bigvee_i^j$ and $\bigwedge_i^j$ to denote sum and product of symbols respectively. For example, $\bigvee_{i=0}^2 a_i$ implies $(a_0 \vee a_1 \vee a_2)$, and $\bigwedge_{i=0}^2 a_i$ implies $(a_0 \wedge a_1 \wedge a_2)$.

Modeling Conditions in FSM

Let us assume that every instruction register $IR_{i,j}$ has an exception bit $XN_{IR_{i,j}}$, which is set when the exception condition ($cond_{IR_{i,j}}^{XN}$) is true. The $XN_{IR_{i,j}}$ has two components viz., exception condition when the children are in exception ($XN_{IR_{i,j}}^{child}$), and exception condition due to exceptions on $IR_{i,j}$ ($XN_{IR_{i,j}}^{self}$). More formally the exception condition at time t in the presence of a set of external signals $I(t)$ on $S(t)$ is, $cond_{IR_{i,j}}^{XN}(S(t), I(t))$, $cond_{IR_{i,j}}^{XN}$ in short,

$$cond_{IR_{i,j}}^{XN} = XN_{IR_{i,j}} = XN_{IR_{i,j}}^{child} \vee XN_{IR_{i,j}}^{self} \quad (3.6)$$

For example, if the designer specified that *any* (see Section 3.2.1(A)) of the children is responsible for the exception on $IR_{i,j}$ i.e., $IR_{i,j}$ will be in exception condition if any of its children is in exception, the Equation (3.6) becomes:

$$XN_{IR_{i,j}} = (\bigvee_{k=1}^{q} XN_{IR_{i+1,k}}) \vee XN_{IR_{i,j}}^{self}$$

Similarly, the conditions for squashing ($cond_{IR_{i,j}}^{SQ}$), stalling ($cond_{IR_{i,j}}^{ST}$), normal flow ($cond_{IR_{i,j}}^{NF}$) and bubble insertion ($cond_{IR_{i,j}}^{BI}$) are shown below:

$$cond_{IR_{i,j}}^{SQ} = SQ_{IR_{i,j}} = NF_{IR_{i,j}}^{parent} \wedge ST_{IR_{i,j}}^{child} \wedge ((IR_{i,j}).opcode == nop) \quad (3.7)$$

$$cond_{IR_{i,j}}^{ST} = (ST_{IR_{i,j}}^{child} \vee ST_{IR_{i,j}}^{self}) \wedge \overline{XN_{IR_{i,j}}} \wedge \overline{SQ_{IR_{i,j}}} \quad (3.8)$$

$$cond_{IR_{i,j}}^{NF} = NF_{IR_{i,j}}^{parent} \wedge NF_{IR_{i,j}}^{child} \wedge \overline{ST_{IR_{i,j}}^{self}} \wedge \overline{XN_{IR_{i,j}}} \wedge \overline{SQ_{IR_{i,j}}} \quad (3.9)$$

$$cond_{IR_{i,j}}^{BI} = BI_{IR_{i,j}}^{parent} \wedge BI_{IR_{i,j}}^{child} \wedge \overline{ST_{IR_{i,j}}^{self}} \wedge \overline{XN_{IR_{i,j}}} \wedge \overline{SQ_{IR_{i,j}}} \quad (3.10)$$

Similarly the conditions for PC viz., $cond_{PC}^{SE}$ (SE: sequential execution), $cond_{PC}^{BI}$ (BI: bubble insertion), and $cond_{PC}^{BT}$ (BT: branch taken) can be described using the information available in the ADL. The $cond_{PC}^{BT}$ is true when a branch is taken or when an exception is taken. When a branch is taken, the PC is modified with the target address. When an exception is taken, the PC is updated with the corresponding interrupt service routine address. Let us assume that the BT_{PC} bit is set when the unit completes execution of a branch instruction and the branch is taken. Formally,

$$cond_{PC}^{SE} = NF_{PC}^{child} \wedge \overline{ST_{PC}^{self}} \wedge \overline{BT_{PC}} \wedge \overline{XN_{IR_{1,j}}} \tag{3.11}$$

$$cond_{PC}^{ST} = (ST_{PC}^{child} \vee ST_{PC}^{self}) \wedge \overline{BT_{PC}} \wedge \overline{XN_{IR_{1,j}}} \tag{3.12}$$

$$cond_{PC}^{BT} = (BT_{PC} \vee XN_{IR_{1,j}}) \tag{3.13}$$

Modeling State Transition Functions

In this section, we describe the next-state function of the FSM. Figure 3.9 shows a fragment of the processor pipeline with only instruction registers. If there are no pipeline hazards, instructions flow from IR (instruction register) to IR every m cycles (m = 1 for single-cycle IR). In this case, the instruction in $IR_{i-1,l}$ $(1 \leq l \leq p)$ at time t proceeds to $IR_{i,j}$ after m cycles i.e., $IR_{i,j}(t+1) = IR_{i-1,l}(t)$. In the presence of pipeline hazards, however, the instruction in $IR_{i,j}$ may be stalled, i.e., $IR_{i,j}(t+1) = IR_{i,j}(t)$. Note that, in general, any instruction in the pipeline cannot skip pipeline stages. For example, $IR_{i,j}(t+1)$ cannot be $IR_{i-2,v}(t)$ $(1 \leq v \leq n_{i-2})$ if there are no feed-forward paths.

The rest of this section formally describes the next-state function of the FSM. According to the Equation (3.5), a state of a $n-$stage pipeline is defined by $(M+1)$ registers (PC and M instruction registers where, $M = \sum_{i=1}^{n-1} n_i$). Therefore, the next state function of the pipeline can also be decomposed into $(M+1)$ sub-functions each of which is dedicated to a specific state register. Let f_{PC}^{NS} and $f_{IR_{i,j}}^{NS}$ $(1 \leq i \leq n-1,\ 1 \leq j \leq n_i)$ denote next-state functions for PC and $IR_{i,j}$ respectively. Note that in general $f_{IR_{i,j}}^{NS}$ is a function of not only $IR_{i,j}$ but also other state registers and external signals from outside of the controller. For the program counter, we define three types of state transitions as follows:

$$\begin{aligned} & PC(t+1) \\ = \ & f_{PC}^{NS}(S(t), I(t)) \\ = \ & \begin{cases} PC(t) + L & \text{if } cond_{PC}^{SE} = 1 \\ target & \text{if } cond_{PC}^{BT} = 1 \\ PC(t) & \text{if } cond_{PC}^{ST} = 1 \end{cases} \end{aligned} \tag{3.14}$$

Here, $I(t)$ represents a set of external signals at time t, L represents the instruction length, and *target* represents the branch target address. The conditions ($cond_{PC}^{x}$, $x \in SE, BT, ST$) are logic functions of $S(t)$ and $I(t)$ as described in Equation (3.11) - Equation (3.13), and return either 0 or 1. For example, if $cond_{PC}^{ST}$ is 1, PC keeps its current value at the next cycle.

For instruction registers, $IR_{i,j}$ ($2 \leq i \leq n-1$, $1 \leq j \leq n_i$), we define five types of state transitions as follows. The state transitions for the first instruction register, $IR_{1,j}$, will have $IM(PC(t))$ in place of $IR_{i-1,l}(t)$, where $IM(PC(t))$ denotes the instruction pointed by the program counter (PC) in instruction memory (IM).

$$\begin{aligned} & IR_{i,j}(t+1) \\ = \; & f_{i,j}^{NS}(S(t), I(t)) \\ = \; & \begin{cases} IR_{i-1,l}(t) & \text{if } cond_{IR_{i,j}}^{NF} = 1 \\ IR_{i,j}(t) & \text{if } cond_{IR_{i,j}}^{ST} = 1 \\ nop & \text{if } cond_{IR_{i,j}}^{BI} = 1 \\ IR_{i-1,l}(t) & \text{if } cond_{IR_{i,j}}^{SQ} = 1 \\ nop & \text{if } cond_{IR_{i,j}}^{XN} = 1 \end{cases} \end{aligned} \tag{3.15}$$

The $IR_{i,j}$ is said to be stalled at time t if $cond_{IR_{i,j}}^{ST}$ is 1, resulting in $IR_{i,j}(t+1) = IR_{i,j}(t)$. Similarly, $IR_{i,j}$ is said to flow normally at time t if $cond_{IR_{i,j}}^{NF}$ is 1. A *nop* instruction (bubble) is inserted in $IR_{i,j}$ when $cond_{IR_{i,j}}^{BI}$ or $cond_{IR_{i,j}}^{XN}$ is 1, resulting in $IR_{i,j}(t+1) = nop$. Similarly, when $cond_{IR_{i,j}}^{SQ}$ is 1, the bubble in $IR_{i,j}$ gets overwritten by the instruction from the parent instruction register, i.e., $IR_{i,j}(t+1) = IR_{i-1,l}(t)$ $(1 \leq l \leq n_{i-1})$.

In this FSM model, signals coming from the datapath or the memory subsystem into the pipeline controller are modeled as primary inputs to the FSM, and control signals to the datapath or the memory subsystem are modeled as outputs from the FSM.

3.2.2 Validation of Dynamic Properties

Based on the FSM model presented in Section 3.2.1, we propose a method for validating dynamic behaviors of pipelined processor specifications using two properties: determinism and in-order execution. We consider validation of dynamic behavior for architectures with in-order execution. We first describe the properties needed for validating the specification. Next, we present an automatic property checking framework driven by the EXPRESSION ADL [5].

A. Properties

This section presents two properties: determinism and in-order execution. Any pipelined processor with in-order execution must satisfy these properties.

Determinism

To ensure correct execution, there should not be any instruction or data loss in the pipeline. The bubble squashing and flushing of instructions are permitted. The flushed instructions are fetched and executed again. The next-state functions for all state registers must be deterministic. This property is valid if all the following equations hold for $\forall i, j(1 \leq i \leq n-1, 1 \leq j \leq n_i)$.

$$cond_{PC}^{SE} \vee cond_{PC}^{BT} \vee cond_{PC}^{ST} = 1 \tag{3.16}$$

$$cond_{IR_{i,j}}^{NF} \vee cond_{IR_{i,j}}^{ST} \vee cond_{IR_{i,j}}^{BI} \vee cond_{IR_{i,j}}^{XN} \vee cond_{IR_{i,j}}^{SQ} = 1 \tag{3.17}$$

$$\forall x, y(x, y \in \{SE, BT, ST\} \wedge x \neq y),\ cond_{PC}^{x} \wedge cond_{PC}^{y} = 0 \tag{3.18}$$

$$\forall x, y(x, y \in \{NF, ST, BI, XN, SQ\} \wedge x \neq y),\ cond_{IR_{i,j}}^{x} \wedge cond_{IR_{i,j}}^{y} = 0 \tag{3.19}$$

The first two equations mean that in the next-state function for each state register, the five conditions must cover all possible combinations of processor states $S(t)$ and external signals $I(t)$. The last two guarantee that any two conditions are disjoint for each next-state function. Informally, exactly one of the conditions should be true in a clock cycle for each state register. As a result, at any time t an instruction register will have a deterministic instruction. In other words, given an initial state of the pipelined processor and an input application program consisting of instruction sequences, it is possible to deterministically decide the instruction in a given instruction register at a given time t.

In-Order Execution

A pipelined processor with in-order execution is correct if all instructions that are fetched from instruction memory, flow from the first stage to the last stage, while maintaining their execution order. In order to guarantee in-order execution, state transitions of adjacent instruction registers must depend on each other. Illegal combination of state transitions of adjacent stages are described below using Figure 3.9 where $2 \leq i \leq n-1$, $1 \leq j \leq n_i$, $1 \leq l \leq p$, and $1 \leq k \leq q$.

An instruction register can not be in normal flow if all the parent instruction registers (adjacent ones) are stalled. If such a combination of state transitions are allowed, the instruction stored in $IR_{i-1,l}$ at time t will be duplicated, and stored into both $IR_{i-1,l}$ and $IR_{i,j}$ in the next cycle. Therefore, the instruction will be executed more than once. Formally, the Equation (3.20) should be satisfied. Similarly, the remaining equations (Equation (3.21) - Equation (3.32)) should be satisfied for $IR_{i,j}$. The detailed explanation is available in [95].

$$(\bigwedge_{l=1}^{p} cond^{ST}_{IR_{i-1,l}}) \wedge cond^{NF}_{IR_{i,j}} = 0 \tag{3.20}$$

$$cond^{NF}_{IR_{i,j}} \wedge (\bigwedge_{k=1}^{q} cond^{ST}_{IR_{i+1,k}}) = 0 \tag{3.21}$$

$$cond^{BI}_{IR_{i,j}} \wedge (\bigwedge_{k=1}^{q} cond^{ST}_{IR_{i+1,k}}) = 0 \tag{3.22}$$

$$cond^{NF}_{IR_{i-1,l}} \wedge cond^{BI}_{IR_{i,j}} = 0 \tag{3.23}$$

$$cond^{BI}_{IR_{i-1,l}} \wedge cond^{BI}_{IR_{i,j}} = 0 \tag{3.24}$$

$$cond^{ST}_{IR_{i-1,l}} \wedge cond^{SQ}_{IR_{i,j}} = 0 \tag{3.25}$$

$$cond^{XN}_{IR_{i-1,l}} \wedge cond^{SQ}_{IR_{i,j}} = 0 \tag{3.26}$$

$$cond^{SQ}_{IR_{i,j}} \wedge cond^{NF}_{IR_{i+1,k}} = 0 \tag{3.27}$$

$$cond^{SQ}_{IR_{i,j}} \wedge cond^{NI}_{IR_{i+1,k}} = 0 \tag{3.28}$$

$$cond^{NF}_{IR_{i-1,l}} \wedge cond^{XN}_{IR_{i,j}} = 0 \tag{3.29}$$

$$cond^{ST}_{IR_{i-1,l}} \wedge cond^{XN}_{IR_{i,j}} = 0 \tag{3.30}$$

$$cond^{SQ}_{IR_{i-1,l}} \wedge cond^{XN}_{IR_{i,j}} = 0 \tag{3.31}$$

$$cond^{BI}_{IR_{i-1,l}} \wedge cond^{XN}_{IR_{i,j}} = 0 \tag{3.32}$$

The above equations are not sufficient to ensure in-order execution in fragmented pipelines. An instruction I_a should not reach join node earlier than an instruction I_b when I_a is issued by the corresponding fork node later than I_b. Formally the following equation should hold:

$$\forall (F,J), I_a \preceq_J I_b \Rightarrow \Gamma_F(I_a) < \Gamma_F(I_b) \tag{3.33}$$

where, (F, J) is fork-join pair, $I_a \preceq_J I_b$ implies I_a reached join node J before I_b, $\Gamma_F(I_a)$ and $\Gamma_F(I_b)$ returns the timestamps when instructions I_a and I_b (respectively) are issued by the fork node F.

The previous property ensures that instruction does not execute out-of-order. However, with the current modeling, two instructions with different timestamps can reach the join node. If join node does not have capacity for more than one instruction, this may cause instruction loss. We need the following property to ensure that only one immediate parent of the join node is in normal flow at time t:

$$\forall x,y(x,y \in \{1,2,...,p\} \wedge x \neq y),\ cond^{NF}_{IR_{i-1,x}} \wedge cond^{NF}_{IR_{i-1,y}} = 0 \quad (3.34)$$

Similarly, the state transition of PC must depend on the state transition of $IR_{1,j}$ $(1 \leq j \leq n_1)$. The illegal combination of state transitions between PC and $IR_{1,j}$ are described below:

$$cond^{ST}_{PC} \wedge cond^{NF}_{IR_{1,j}} = 0 \quad (3.35)$$

$$cond^{SE}_{PC} \wedge (\bigwedge_{j=1}^{n_1} cond^{ST}_{IR_{1,j}}) = 0 \quad (3.36)$$

$$cond^{BT}_{PC} \wedge (\bigwedge_{j=1}^{n_1} cond^{ST}_{IR_{1,j}}) = 0 \quad (3.37)$$

$$cond^{SE}_{PC} \wedge cond^{BI}_{IR_{1,j}} = 0 \quad (3.38)$$

$$cond^{BT}_{PC} \wedge cond^{BI}_{IR_{1,j}} = 0 \quad (3.39)$$

$$cond^{SE}_{PC} \wedge cond^{XN}_{IR_{1,j}} = 0 \quad (3.40)$$

$$cond^{ST}_{PC} \wedge cond^{SQ}_{IR_{1,j}} = 0 \quad (3.41)$$

$$cond^{ST}_{PC} \wedge cond^{XN}_{IR_{1,j}} = 0 \quad (3.42)$$

We have described all possible illegal combination of state transition functions (Equation (3.20) - Equation (3.42)). However, Equation (3.23), Equation (3.24), Equation (3.27), and Equation (3.28) are not necessary to prove in-order execution.

B. Automatic Validation Framework

Algorithm 6 describes the specification validation technique. It accepts the processor specification as input. The FSM model and the properties are generated from the ADL specification. In case of a failure, it generates counter-examples so that the designer can modify the ADL specification of the architecture.

We have verified the properties using two different approaches. First, we have used an SMV [43] based property checking framework as shown in Figure 3.10. The SMV based approach fits nicely in our validation framework. However, the SMV is limited by the size of the design it can handle. We have also developed an equation solver based framework as shown in Figure 3.11 that can handle complex designs. In this section, we briefly describe these two approaches. The detailed description is available in [95].

Validation using Model Checker

The FSM model (SMV description) of the processor is generated from the ADL specification. The properties are also described using SMV description. The properties are applied on the FSM model using the SMV model checker as shown in Figure 3.10. In case of failure, SMV generates counter-examples that can be used to modify the ADL specification. Each counter-example describes the failed equation(s) and the instruction registers that are involved.

Algorithm 6: *Validation of Pipeline Specification*
Input: ADL specification of the processor architecture.
Outputs: *Success*, if the processor model satisfies the properties.
Failure otherwise, and produces the counter-examples.

```
{
  Generate FSM model using Equation (3.5) - Equation (3.15)
  Generate properties using Equation (3.16) - Equation (3.42)
  Apply the properties on the FSM model.
  Return Success if all the properties are verified;
         Failure otherwise, and produce the counter-example(s).
}
```

We have verified the in-order execution style of the processor specification in two ways. First, the framework generates properties using Equation (3.20) - Equation (3.42) to verify in-order execution. This is similar to how other properties (e.g., determinism) are verified. Second, an auxiliary automata is used instead of using equations to verify in-order execution.

In the auxiliary automata based approach, we use the same FSM model of the processor (SMV description) generated from the ADL specification. We have developed a SMV module that generates two instructions randomly with random delay between them. These two instructions are recorded and fed to the FSM model. The processor (FSM) model accepts these instructions and performs regular computations. At the completion (e.g., writeback unit) the auxiliary automata analyzes these two instructions to see whether they completed in the same sequence as generated. Note that, this auxiliary automata does not need any manual modification for different architectures. In case of failure, SMV generates counter-examples containing instruction sequences (instruction pair with NOPs in between them) that violate in-order execution for the processor model.

Validation using Equation Solver

In the second approach, the framework generates the FSM model and flow equations for each instruction register and PC using ADL specification and Equation (3.5) - Equation (3.15). The framework generates the equations necessary for verifying properties using ADL description and Equation (3.16) - Equation (3.42) as shown in Figure 3.11.

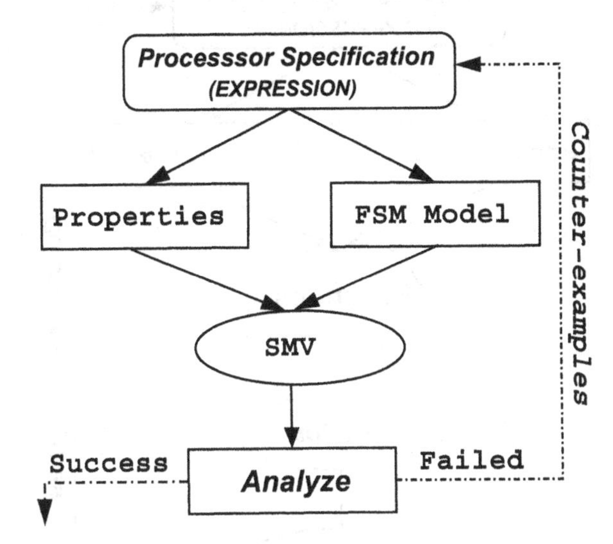

Figure 3.10: Automatic validation framework using SMV

The *Eqntott* [41] tool converts these equations in two-level representation of a two-valued Boolean function. This two-level representation is fed to *Espresso* [40] tool that produces minimal equivalent representation. Finally, the minimized representation is analyzed to determine whether the property is successfully verified or not. In case of failure, it generates traces explaining the cause of failure. The trace contains the equation(s) that failed, and the identification of the instruction registers involved. The designer therefore knows the property that is violated and the reason for the violation. This information is used to modify the ADL specification.

3.2.3 A Case Study

In a case study we successfully applied the proposed methodology to the single-issue DLX [55] processor. We used the EXPRESSION ADL [5] to capture the structure and behavior of the DLX processor shown in Figure 3.6. We captured the conditions for stalling, normal flow, exception, branch taken, squashing, and bubble insertion in the ADL. Using the ADL description, we automatically generated the equations for flow conditions for all the units. The necessary equations for verifying the properties such as determinism and in-order execution are generated automatically from the given ADL specification. The detailed description of the case study is available in Appendix D.

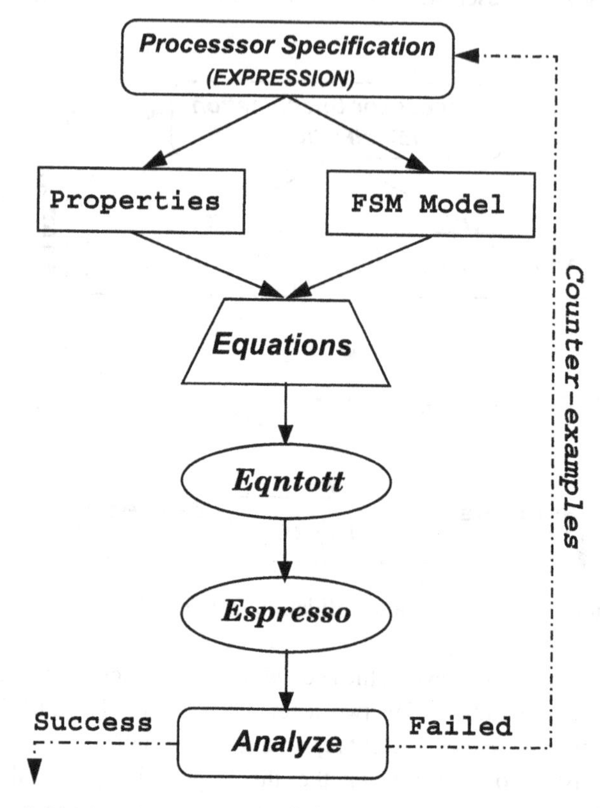

Figure 3.11: Automatic validation framework using equation solver

We have verified the properties using two different methods: using the *SMV* model checker and the *Espresso* equation solver, as described in Section 3.2.2. We have used a 300 MHz Sun UltraSparc-II with 1024M RAM to run the experiments. Table 3.3 shows the performance of the two methods for verifying in-order execution property. We have used the VLIW DLX architecture as the base configuration and modified the number of opcodes. The first column presents our two methods of specification validation. The second, third, and fourth columns present the execution time (in seconds) of the two methods for verifying in-order execution property for different architecture configurations.

Table 3.3: Validation of in-order execution by two frameworks

	DLX Processor Configurations		
	8 opcodes	16 opcodes	32 opcodes
SMV based Framework	302.4 sec	400.4 sec	740.9 sec
Espresso based Framework	5.4 sec	6.7 sec	9.4 sec

We have performed experiments by modifying the pipeline structure such as addition of pipeline paths and pipeline stages. Our SMV based framework could not verify in-order execution when pipeline path is added to the VLIW DLX architecture. However, our equation solver based framework can handle complex configurations. The SMV based framework performed better for verifying the determinism. This is due to the fact that the properties (equations) that need to be applied to verify determinism consists of local computations for each state register. The SMV based framework took 0.8 seconds to verify determinism property, whereas the equation solver based framework took 4 seconds for the same DLX configuration. Although we have not applied this technique on other architectures, we believe the SMV based framework is suitable for verifying the determinism property whereas our equation solver based framework can be used for verifying in-order execution of complex architectures.

3.3 Related Work

The problem of verifying a given specification has been studied extensively for hardware as well as software designs. Although different specification languages are used depending on the level of granularity and expressiveness needed to specify the target design, the specification verification is performed typically in two ways: property checking and simulation. First, the specification is analyzed to ensure that it satisfies a set of necessary properties. Second, the executable specification is simulated using a set of test vectors and the generated outputs are compared with the expected results. The input test vectors can be generated from the specification as well [87].

Verification of design specification has two major applications: verification of requirements specification during software development, and verification of protocols in both software and hardware specifications. The verification activities that accompany requirements stage of software development ensure the adequacy of the requirements (including correctness, completeness, and consistency), and generate the initial testcases with the expected (correct) responses [136].

There has been a plethora of previous work in the area of protocol verification. Bunker et al. [2] surveyed formal hardware specification languages for protocol compliance verification. A comprehensive survey of various approaches for the verification of cache coherence protocols based on state enumeration, (symbolic) model checking, and symbolic state models is presented by Pong et al. [29]. Shimizu et al. presented techniques to specify and verify commonly used interface protocols such as PCI bus protocol and Intel Itanium bus protocol [67].

3.4 Chapter Summary

Validation of the architectural specification is essential to ensure that the reference model is golden so that it can be used to uncover bugs in the design. This chapter presented a framework for automatic modeling and validation of pipelined processor specifications driven by an ADL.

We developed validation techniques to ensure that the static behavior of the pipeline is well-formed by analyzing the structural aspects of the specification using a graph based model. We applied these techniques on the graph model of the MIPS R10K, TI C6x, DLX, and PowerPC architectures to demonstrate the usefulness of this approach. The dynamic behavior is verified by analyzing the instruction flow in the pipeline using a FSM-based model to validate several important architectural properties such as determinism and in-order execution in the presence of hazards and multiple exceptions. We applied this methodology to the DLX processor to demonstrate the usefulness of this technique.

These properties are by no means complete to prove the correctness of the specification. The designer can add new architecture-specific properties and easily integrate it in our framework. Our validation framework uses two approaches: SMV based property checking and Espresso based equation minimization. The validation framework determines whether all the necessary properties are satisfied. In case of a failure, it generates traces so that a designer can modify the ADL specification of the architecture.

Part III

Top-Down Validation

4

EXECUTABLE MODEL GENERATION

Contemporary processor architectures vary widely in terms of their architectural features. Program address generation and instruction dispatch features are widely used in DSP processors. VLIW processors use strong compiler support to ensure correct execution of long instruction words. Superscalar processors on the other hand, use hardware scheduling techniques, register renaming, and so on. Multimedia processors support SIMD operations. Furthermore, each architecture has different branch prediction schemes, execution style (e.g., in-order, out-of-order), interrupt handling procedures, and last but not the least different memory subsystems. Emerging architectures have combined features of classical architectures. For example, the Intel Itanium combines the features of VLIW and superscalar architectures; the TI C6x family combines the features of DSP and VLIW architectures. Designers of customized programmable embedded architectures need the ability to explore and evaluate a variety of heterogeneous processor-memory architectures, and thus need a framework that can capture a wide range of such architectural features; such a framework will facilitate rapid design space exploration of heterogeneous processor-memory architectures.

Moreover, during design space exploration using customized Intellectual Property (IP) cores designers may want to add certain architectural features (e.g., some superscalar features to a VLIW processor core) to see how it impacts the area, power, performance, and other important design parameters. Similarly, to find the best match between the application characteristics and the memory organization features (e.g., caches, stream buffers, access modes, SRAM, DRAM etc.), the designer needs to explore different memory configurations in combination with different processor architectures, and evaluate each such system for cost, power, and performance. To enable this, designers need (i) a way of specifying a wide variety of processor-memory features, and (ii) the ability to generate automatically executable models of the architecture. Functional abstraction techniques are essential for this purpose: it is necessary to find the common basis among such heterogeneous architectures and use that as a building block for defining a set of abstraction primitives. The abstraction primitives should be simple enough to allow correlation with the architectural features. On the other hand, the primitives should be generic enough to be useful across a wide range of architectures. In this chapter we present a functional abstraction technique that enables automatic generation of executable models from the ADL specification.

This chapter is organized as follows. Section 4.1 surveys contemporary programmable architectures. Section 4.2 presents the functional abstraction needed to capture a wide variety of architectural features and memory configurations. Section 4.3 describes the procedure for reference model generation from the ADL specification using functional abstraction followed by the related work in Section 4.4. Finally, Section 4.5 summarizes the chapter.

4.1 Survey of Contemporary Architectures

We have studied contemporary processor and memory architectures from popular architectural domains [97]. This section summarizes the survey and outlines the similarities and differences of the architectural features available in a wide a variety of processor and memory architectures.

4.1.1 Summary of Architectures Studied

In order to understand and characterize the diversity of contemporary architectures, we have surveyed processors from different architectural domains - RISC (MIPS R4000 [81] and StrongArm [125]), DSP (Motorola 56000 and TI C5x), VLIW DSP (TI C6x [131], MAP1000A [16], and Motorola StarCore [47]), superscalar (MIPS R10000 [50], MPC7450 [48], Sun UltraSparc IIi [126], and DEC Alpha 21364), and hybrid (Intel IA-64 [138]). The Intel IA-64 architecture has combined features of VLIW and superscalar processors with out-of-order execution. Table 4.1 summarizes the processor-memory features for different architectures. Each row of the table corresponds to an architectural feature. Each column represents an architecture. We have shown only the relationship between a feature and an architecture. In general, an architectural feature may also depend on the type of the instruction.

An entry in Table 4.1, *TAB[F, A]*, represents the behavior of an architecture *"A"* towards a feature *"F"*. If an entry is marked *x*, that feature is supported by that architecture. If an entry is blank, the feature is either not supported or not applicable (or not known) for that architecture. An entry containing an integer number, *n*, indicates that the feature is supported *n* times. An entry containing a series, *(n-m)*,

Architectures	**RISC**		**DSP**		**VLIW DSP**			**Superscalar**				**Hybrid**
Processor-Memory Features	R4K	SA	56K	C5x	C6x	MA	SC	R10	MP	U3	α64	IA64
# of fetches/cycle	2	1	1	1	8	4	8	4	4	4	4	6
# of fetch stages	2	1	1	1	4		3	1	2	1	1	2
# of decodes/cycle	2	1	1	1	8	4		4	3	4	4	
# entries in decode RS									12			8
# of issue units								3	3	1	3	3
# of issues/cycle							6	5	6	4	6	6
# entries in issue RS								48	12		35	
# operations/instruction	1	1	1	1	8	4		1	1	1	1	
# of parallel exec units					8	4	6	5	11		6	
Branch Prediction								2b	BT		MA	
Feedback paths		x								x		x
Operand read in	D	D	E	R	E	E	E	I	I	I	R	I
SIMD support						x			x	x		x
entries in completion Q								32	16			
Register Renaming								x	x		x	x
Dynamic Scheduling								x	x	x	x	x
Speculation											x	x
Predication						x						x
# register files	2	1	3	1	3	3	2	2	3		3	5
# Coprocessors	3	1										
# pipeline stages	8	5	3	4	3		5	5-7	7	9	6	10
Levels of D-Cache	1-2	1			0-2	1	0-2	2	3	2	2	3
cache prefetch		x								x		x
cache hints												x
On-chip SRAM		x			x	x	x			x		
configurable SRAM					x							
Off-chip DRAM	x	x	x	x	x	x	x	x	x	x	x	x
page/burst mode		x			x							
Write Buffer		x								x	x	
Read Buffer		x										
Victim Buffer											x	
Stack			x	x			x					
FIFO						x						
DMA		x			x	x	x				x	
parallel mem transfers	1		2		2		2	1			2	2
mem pipelining								x			x	

Table 4.1: Processor-memory features of different architectures. *R4K: MIPS R4000, SA: StrongArm, 56K: Motorola 56K, c5x: TI C5x, c6x: TI C6x, MA: MAP1000A, SC: Starcore, R10: MIPS R10000, MP: Motorola MPC7450, U3: SUN UltraSparc IIi, α64: Alpha 21364, IA64: Intel IA-64*

implies that the feature is supported for i times, where $(n <= i <= m)$. Similarly, an entry containing a set, $\{n,m\}$, means that the feature is supported either n or m times. For example, the table entry with memory feature "*Levels of D-Cache*" and processor name *IA64* has value 3; it implies that IA-64 has 3 levels of data cache. The row corresponding to "*operand read in*" has four types of values depending on where the operands are read in the pipeline: (D: Decode stage), (R: Read stage), (I: Issue stage), and (E: Execute stage). The row corresponding to *Branch Prediction* has values that indicate the method of branch prediction employed in the respective architecture: (2b: 2-bit algorithm using branch history table), (BT: BTB based prediction), and (MA: dynamically choose among multiple algorithms based on local predictor table, global predictor table and branch history table).

4.1.2 Similarities and Differences

Broadly speaking, the structure of a processor consists of functional units (such as fetch, decode, issue etc.), connected using ports, connections and pipeline latches. Similarly, the structure of a memory subsystem consists of SRAM, DRAM, cache hierarchy, and so on. Although a broad classification makes the architectures look similar, each architecture differs in terms of the algorithm it employs in branch prediction, the way it detects hazards, the way it handles exceptions etc. Moreover, each unit has different parameters for different architectures (e.g., number of fetches per cycle, levels of cache, cache line size etc.).

Depending on the architecture a functional unit may perform the same computation at different stages in the pipeline. For example, read-after-write (RAW) followed by operand read happen in the decode unit for some architectures (e.g., DLX [55]), whereas in some others these operations are performed in the issue unit (e.g., MIPS R10K [50]). Some architectures even allow operand read in the execution unit. On the other hand, some architectures do not issue operations if RAW hazard is detected while others issue the operation in spite of RAW hazard (e.g., use snooping to read the data at execution stage using feedback paths). In other words, the same functionality is used at different stages in the pipeline for different architectures.

Towards obtaining a unifying abstraction, we can observe some fundamental differences from the study; the architecture may use:

1. the same functional or memory unit with different parameters
2. the same functionality in different functional or memory unit
3. new architectural features

The first difference can be eliminated by defining generic functions with appropriate parameters. The second difference can be eliminated by defining generic sub-functions that can be used by different architectures at different stages in the pipeline. The last one is difficult to alleviate since it is new, unless this new functionality can be composed of existing sub-functions. Section 4.2 presents the functional abstraction needed to capture a wide variety of architectural features and memory configurations.

4.2 Functional Abstraction

Functional abstraction allows the system designer to describe a wide variety of architectures. In this section we present functional abstraction by way of illustrative examples. We first explain the functional abstraction needed to capture the structure and behavior of the processor and memory subsystem, then we discuss the issues related to defining a generic controller functionality, and finally we discuss the issues related to handling interrupts and exceptions.

4.2.1 Structure of a Generic Processor

The structure of each functional unit is captured using parameterized functions. However, generic functions are not sufficient since each functional unit may perform a different set of computations depending on the architecture. Hence, there is a need for parametric sub-functions. Based on the observations made in Section 4.1, we have defined a set of common functions and sub-functions with appropriate parameters. First, we describe the generic functions. Next, we describe the generic sub-functions. Finally, we discuss how these functions and sub-functions are used to compose a new processor architecture.

Generic functions

We capture the structure of each functional unit using parameterized functions. For example, a fetch unit functionality contains several parameters, such as number of operations read per cycle, number of operations written per cycle, reservation station size, branch prediction scheme, number of read ports, number of write ports, and so on. Figure 4.1 shows a specific example of a fetch unit described using sub-functions. Each sub-function is defined using appropriate parameters. For example, *ReadInstMemory* reads n operations from instruction cache using current PC address (returned by *ReadPC*) and writes them to the reservation station. The fetch unit reads m operations from the reservation station and writes them to the

```
FetchUnit ( # of read/cycle, res-station size, ....)
{
   address = ReadPC();
   instructions = ReadInstMemory(address, n);
   WriteToReservationStation(instructions, n);
   outInst = ReadFromReservationStation(m);
   WriteLatch(decode_latch, outInst);

   pred = QueryPredictor(address);
   if pred {
      nextPC = QueryBTB(address);
      SetPC(nextPC);
   } else
      IncrementPC(x);
}
```

Figure 4.1: A fetch unit example

output latch (fetch to decode latch) and uses BTB based branch prediction mechanism. We have defined parameterized functions for all functional units present in contemporary programmable architectures [97].

Generic sub-functions

We have defined sub-functions for all the common activities e.g., ReadLatch, WriteLatch, ReadOperand, and so on. Table 4.2 lists some of the common activities that we have identified. The first column represents the name of the function, the second column describes the activity, and the last column describes the input and output parameters of the function. We have also defined a set of sub-functions including *RenameRegister* and *GraduateOperation* using sub-functions. Figure 4.2 shows a specific implementation of *RenameRegister* modeled using sub-functions.

New architecture generation using generic functions and sub-functions

So far we have discussed the generic functions and sub-functions necessary to capture a wide variety of processor architectures. In this section we briefly describe how these functions and sub-functions can be used to compose a new architecture or modify an existing architecture. First, we describe how to compose a simple RISC architecture. Next, we discuss what generic functions are necessary to modify the simple RISC architecture into a VLIW or superscalar architecture.

Table 4.2: A list of common sub-functions

Function Name	Description	Parameters
ReadLatch	Read a latch for n operations	Latch X, n, Data
WriteLatch	Write data to a latch	Latch, Data
QueryPredictor	Query prediction status	Branch address, status
QueryBTB	Query predicted address	Branch and memory address
UpdateBTB	Send address to branch unit	ID, target address
UpdatePredictor	Update branch predictor	ID, prediction type
BranchOther	Other branch address	ID, Address
IncrementPC	Increase PC with X	X, New PC
SetPC	New PC address X	X, New PC
ReadPC	Get PC	PC address
RSInsertOperation	Add one operation to RS	Operation
RSInsertOperations	Add X operations to RS	Operations, X
RSDeleteOperation	Dequeue operation from RS	ID
RSReadOperation	Read one operation from RS	Operation
RSReadOperand	Read n's operations operands	RS, n, RS
ReadOperand	Read one operand	Address bus, Reg name, Data
WriteResult	Write operand	Address bus, Reg name, Data
MarkDestBusy	Mark Register busy	Register name
ReleaseDest	Unmark Register busy	Register name
CheckRAW	Check for RAW	Register name, status
CheckWAW	Check for WAW	ID, status
CheckWAR	Check for WAR	ID, status
IsUnitBusy	Is unit X busy	X, status
IsUnitStalled	Is unit X stalled	X, status
IsOperandRead	Is operand X read	ID, X, status
MarkOperandRead	Mark the operand as read	ID, X
HasUnitRS	Does unit X have RS?	X, status
SetUnitStalled	Set Stall bit for unit X	X, True/False
SetUnitBusy	Set Busy Bit for unit X	X, True/False
ReadPredicate	Check predicate register X	Pred reg. X, status
WritePredicate	Set predicate register X to Y	Pred reg., value
CheckPredicate	Query ID's predicate	ID, status
ExecuteOperation	Execute an operation	Src1, Src2, func, Result
MarkOperationDone	Mark operation done in comp queue	ID
IsOperationDone	Query if operation done	ID, status
CompletionQInsertOper	Insert an entry into comp queue	Operation
CompletionQDeleteOper	Delete an entry from comp queue	ID
FlushCompletionQ	Remove all operations above ID	ID
IsOperationValid	Query if operation is valid	ID, status
SetValidBit	Set valid bit to X for operation	ID, X
IsBranchAhead	Is there a branch ahead?	ID, status
IsBranchOperation	Is operation a branch?	ID, status
IsStoreOperation	Is operation a store?	ID, status
IsMapped	Is X in mapping table	Reg X, status
GetPhysicalRegister	For a logical register	Logical, physical reg
GetFreeRegister	Return a free physical reg	Register number
MapRegisters	logical to physical	Logical, physical reg

```
RenameRegister (Instruction inst)
begin
        RenameReg(inst.src1);
        RenameReg(inst.src2);
        if (inst.opcode == store)
          RenameReg(dest);
        else if (inst.opcode != branch)
              freeReg = GetFreeRegister();
              MapRegisters(freeReg, inst.dest);
              MarkDestBusy(freeReg);
        endif
end
```

(a) Rename registers in an instruction

```
RenameReg (Register src)
begin
        if IsMapped(src)
          reg = GetPhysicalRegister(src)
        else
          reg = GetFreeRegister();
        endif
        MapRegisters(reg,src)
end
```

(b) Rename a register

Figure 4.2: Modeling of RenameRegister function using sub-functions

A RISC architecture typically has four pipeline stages: fetch, decode, execute, and writeback. Each of these stages requires one generic function. Each function uses *ReadLatch* sub-function to read the instruction from the pipeline latch. At the end of the computation each function uses *WriteLatch* sub-function to write the modified instruction into the output latch. The fetch function reads the program counter (PC) value using *ReadPC*. If the architecture supports branch prediction, the fetch function needs to use appropriate sub-functions such as *QueryPredictor*, *QueryBTB*, and so on. Depending on the outcome, the fetch function either invokes *IncrementPC* or *SetPC*. Similarly, the decode function uses *CheckRAW*, *CheckWAR*, and *CheckWAW* sub-functions to perform hazard detection. The source operands of the instruction are read using *ReadOperand* sub-function. The execute function uses *ExecuteOperation* sub-function to execute an operation. Finally, the writeback function uses *WriteResult* sub-function to write the result back into the register file.

To convert the RISC architecture into a VLIW one that can issue m operations per cycle to the n pipeline paths, we need to perform the following modifications to the functions discussed above. The decode function can use a reservation station (instruction buffer). The instruction buffer can be accessed using sub-functions such as *RSInsertOperation*, *RSDeleteOperation*, and so on. The decode function needs to use *IsUnitBusy* and *IsUnitStalled* sub-functions to detect structural hazard and decide where to send the next instruction. Each pipeline path needs to have separate execute functions. In case the architecture supports predicated execution, a set of sub-functions including *ReadPredicate*, *WritePredicate*, and *CheckPredicate* can be used.

To add superscalar features to the existing VLIW architecture, the following modifications need to be done. The decode function can invoke *RenameRegister* to perform register renaming. Several sub-functions such as *GetPhysicalRegister* and *GetFreeRegister* are also useful in this regard. If the intended execution style is out-of-order execution, we need to add a completion queue (in-order buffer) in the architecture. The decode function needs to insert an operation in the queue using *CompletionQInsertOper* before issuing it to the child unit. This is to ensure in-order completion in the presence of out-of-order execution. The writeback function can delete the front operation(s) of the queue using *CompletionQDeleteOper* sub-function. The completion queue can also be used to perform WAW and WAR checks, to flush necessary instructions in the pipeline, to enforce in-order completion of branches and memory writes, or to synchronize events such as completion of all memory writes and pending exceptions.

4.2.2 Behavior of a Generic Processor

The behavior of a generic processor is captured through the definition of operations. Each operation is defined as a function, with a generic set of parameters, that performs the intended functionality. The parameter list includes source and destination operands, and necessary control and data type information. We have defined common sub-functions (generic set) such as ADD, SUB, MUL, and so on [97].

```
ADD (src1, src2) {
   return (src1 + src2);
}
```

```
MUL (src1, src2) {
   return (src1 * src2);
}
```

```
MAC (src1, src2, src3) {
   return ( ADD( MUL(src1, src2), src3) );
}
```

Figure 4.3: Modeling of MAC operation

Given a new (target) operation and a mapping between the target operation and the generic operations, the functionality of the new operation can be created using the functionalities of the existing operations. For example, the MAC (multiply and accumulate) functionality can be composed of two sub-functions (ADD and MUL) as shown in Figure 4.3.

```
Cache(cache size, line size, ... opType, addr, data)
begin
    // It has three storages: tag, cache, valid
    // Get row, col, and tag from addr
    if opType is READ
      if ((tag[row] == tag) and valid[row])
        data = cache[row][col]
        return HIT
      else
        return MISS
      endif
    else if opType is WRITE
      .........
    else if opType is REPLACE
      .........
    else if opType is REFILL
      .........
    endif
end
```

(a) Cache function

```
AssociativeCache (..., assoc, opType, addr, dataOut)
begin
      if opType is READ
        /** Find the one with data **/
        for ( ci=0; ci < associativity; ci ++)
            stat = Cache(cache_ci, ... READ, data)
            if stat is HIT
              dataOut = data
              return HIT
            endif
        endfor
        // Find the cache to be replaced and refilled
        Cache (..., cache, ..., REPLACE, addr)
        Cache (..., cache, ..., REFILL, addr)
        ......
      else if opType is WRITE
        ......
      endif
end
```

(b) Associative cache function

Figure 4.4: Modeling of associative cache function using sub-functions

4.2.3 Structure of a Generic Memory Subsystem

Each type of memory module, such as SRAM, cache, DRAM, SDRAM, stream buffer, and victim cache, is modeled using a function with appropriate parameters. For example, the cache function shown in Figure 4.4(a) has many parameters including cache size, line size, associativity, word size, replacement policy, write policy, and latency. It performs four operations: read, write, replace, and refill. These functions can have parameters for specifying pipelining, parallelism, access modes (normal read, page mode read, and burst read), and so on. Again, each function is composed of sub-functions. For example, the associative cache function shown in Figure 4.4(b) is modeled using cache sub-function.

4.2.4 Generic Controller

A major challenge in defining an architectural abstraction is the modeling of control for a wide range of architectural styles. We define control in both distributed and centralized manner. The distributed control is transfered through pipeline latches. While an instruction gets decoded the control information needed to select the operation, the source and the destination operands are placed in the output latch as shown in Figure 4.5. These decoded control signals pass through the latches between two pipeline stages unless they become redundant. For example, when the value for *src1* is read that particular control is not needed any more, instead the read value will be in the latch. We have shown here only the control information of the latch. The latch also contains data values and predicate registers (if applicable).

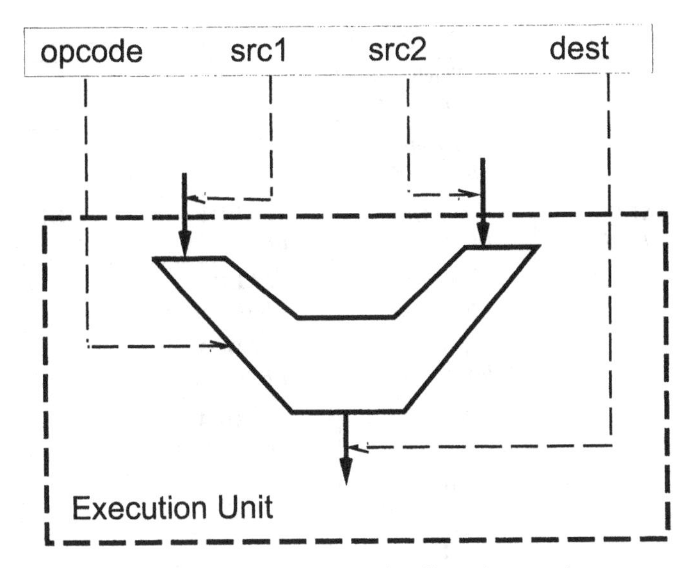

Figure 4.5: Example of distributed control

The centralized control is maintained by using a generic control table. The number of rows in the table is equal to the number of pipeline stages in the architecture. The number of columns is equal to the maximum number of parallel units present in any pipeline stage. Each entry in the control table corresponds to one particular unit in the architecture. It contains information specific to that unit e.g., busy bit (BB), stall bit (SB), list of children, list of parents, opcodes supported, and so on. For example, Figure 4.6 shows the control table for the DLX processor shown in Figure 3.6. The control table captures all the necessary details to perform necessary stalling and flushing of the pipeline.

4.2.5 Interrupts and Exceptions

Another major challenge in defining architectural abstractions is the modeling of interrupts and exceptions. We briefly describe the abstraction needed to capture a wide variety of exceptions and interrupts in programmable architectures. Each exception is captured using an appropriate sub-function. Opcode related exceptions (e.g., divide by zero) are captured in the opcode functionality. Functional unit related exceptions (e.g., illegal slot exception) are captured in functional units. External interrupts (e.g., reset, debug exceptions) are captured in the control unit functionality. Appendix C describes how to capture exceptions and interrupts in an ADL.

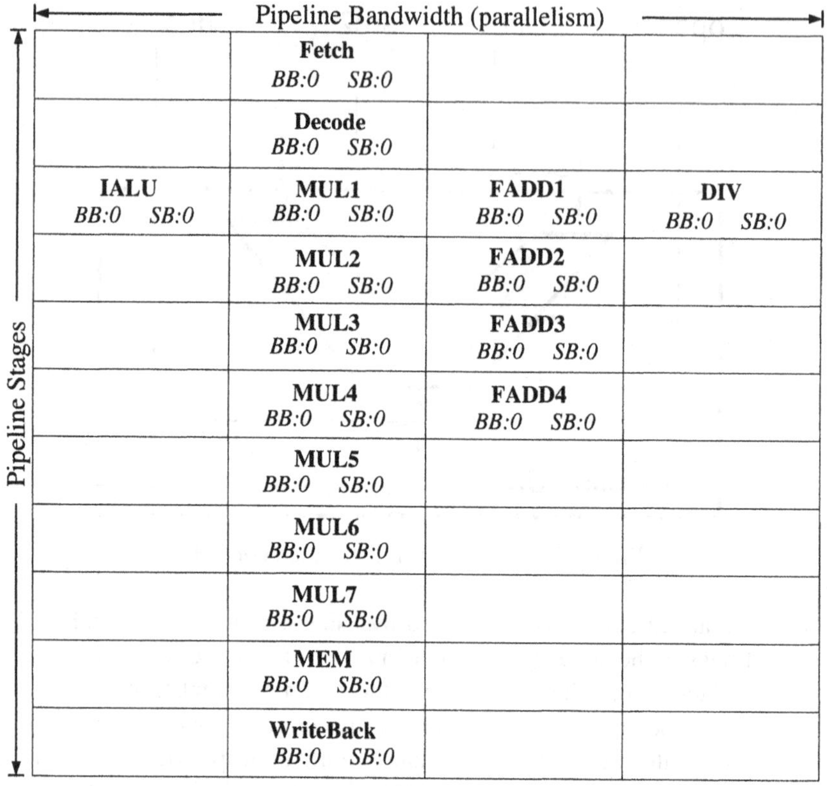

Pipeline Bandwidth (parallelism)

Pipeline Stages

	Fetch *BB:0 SB:0*		
	Decode *BB:0 SB:0*		
IALU *BB:0 SB:0*	**MUL1** *BB:0 SB:0*	**FADD1** *BB:0 SB:0*	**DIV** *BB:0 SB:0*
	MUL2 *BB:0 SB:0*	**FADD2** *BB:0 SB:0*	
	MUL3 *BB:0 SB:0*	**FADD3** *BB:0 SB:0*	
	MUL4 *BB:0 SB:0*	**FADD4** *BB:0 SB:0*	
	MUL5 *BB:0 SB:0*		
	MUL6 *BB:0 SB:0*		
	MUL7 *BB:0 SB:0*		
	MEM *BB:0 SB:0*		
	WriteBack *BB:0 SB:0*		

Figure 4.6: Example of centralized control

We model an interrupt handler unit that services these exceptions. It has information regarding the priority of interrupts and which exceptions generate what interrupt. The generic interrupt handler has a parameterized priority table. The interrupt handler unit generates one particular interrupt based on the priority. Before execution of an interrupt service routine, context saving and complete/partial flushing occurs. The specific type of flushing is decided by the semantics of the interrupt: complete flushing clears the entire pipeline; partial flushing means flushing only the instructions behind the interrupted instruction and allowing the previous instructions to continue using the program order information available in completion queue. Again, these actions are part of the parametric sub-functions that allow a finer grain of microarchitectural exploration.

4.3 Reference Model Generation

We use the functional abstraction technique to generate executable models (such as simulator and synthesizable hardware) from the ADL specification. The procedure is same for generating the simulator, hardware (synthesizable RTL), as well as validation models. The only difference is that the input library (consisting of generic functions and sub-functions) needs to be implemented using the appropriate language. For example, the generic functions and sub-functions need to be implemented using programming languages such as C/C++ to enable simulator generation. Similarly, to enable hardware generation the generic library needs to be implemented using a synthesizable subset of VHDL/Verilog. The development of the generic library (consisting of implementation of generic functions and sub-functions) is a one-time activity and independent of the architecture.

The reference model generation process consists of three steps. First, the ADL specification is read to gather all the necessary details for the model generation. Second, the functionality of each component is composed using the generic functions and sub-functions. Finally, the structure of the architecture is composed using the structural details. In the remainder of this section we describe last two steps for simulator generation. As mentioned earlier, the procedure remains the same for generation of hardware and validation models.

Component Generation

To compose the functionality of each component, all necessary details (such as parameters and functionality) are extracted from the ADL specification. First, we describe how to generate three major components of the processor: instruction decoder, execution unit, and controller, using the generic functions and sub-functions. Next, we describe how to compose the functionality of new instructions (behavior) using the generic functions.

A generic instruction decoder uses information regarding individual instruction format and opcode mapping for each functional unit to decode a given instruction. The instruction format information is available in the ADL specification. The decoder extracts information regarding the opcode and operands from the input instruction using the instruction format. The mapping section of the ADL captures the information regarding the mapping of opcodes to the functional units. The decoder uses this information to perform/initiate necessary functions (e.g., operand read) and decide where to send the instruction.

To compose an execution unit, it is necessary to instantiate all the operation functionalities (e.g, ADD, SUB etc. for an ALU) supported by that execution unit. The execution unit invokes the appropriate opcode functionality for an incoming operation based on a simple table look-up technique as shown in Figure 4.8. Also, if the execution unit is supposed to read the operands, the appropriate number of operand read functionalities need to be instantiated unless the same read functionality can be shared using multiplexers. Similarly, if the execution unit is supposed to write the data back to register file, the functionality for writing the result needs to be instantiated. The actual implementation of an execute unit might contain many more functionalities such as read latch, write latch, modify reservation station (if applicable), and so on.

The controller is implemented in two parts. First, it generates a centralized controller (using generic controller function with appropriate parameters) that maintains the information regarding each functional unit such as busy, stalled etc. It also computes hazard information based on the list of instructions currently in the pipeline. Based on these bits and the information available in the ADL, it stalls/flushes necessary units in the pipeline. Second, a local controller is maintained at each functional unit in the pipeline. This local controller generates certain control signals and sets necessary bits based on the input instruction. For example, the local controller in an execute unit will activate the add operation if the opcode is *add*, or it will set the busy bit in case of a multi-cycle operation.

```
( TARGET
  ( (MACcc dest src1 src2 src3) )
)
( GENERIC
  ( (MUL temp src1 src2) (ADD dest src3 temp) (RESET CR[2]) )
)
```

Figure 4.7: Mapping between *MACcc* and generic instructions

So far we have discussed composition of the structural components for an architecture. It is also necessary to compose the functionality of new instructions (behavior) using the functionality of existing instructions. The EXPRESSION ADL based framework assumes a generic set of instructions (generic architecture). While describing a new architecture (target architecture) using the ADL, it is necessary to provide the mapping between target instructions and generic instructions. This instruction mapping information is typically used by a compiler during instruction selection. This mapping is also used to generate the functionality for the target (new) instructions using the functionality of the corresponding generic instructions. The scheme allows one-to-many, many-to-one, and many-to-many mappings between generic and target instructions. For example, the *MACcc* instruction shown in Figure 4.7 uses three generic instructions. The first two generic instructions perform the multiply and accumulate. The third instruction clears the carry bit of the control register.

Processor Model Generation

The final implementation is generated by instantiating the components (e.g., fetch, decode, ALU, LdSt, writeback, branch, caches, register files, memories etc.) with appropriate parameters and connecting them using the information available in the ADL. For example, Figure 4.8 shows a portion of the simulation model for the DLX architecture shown in Figure 3.6.

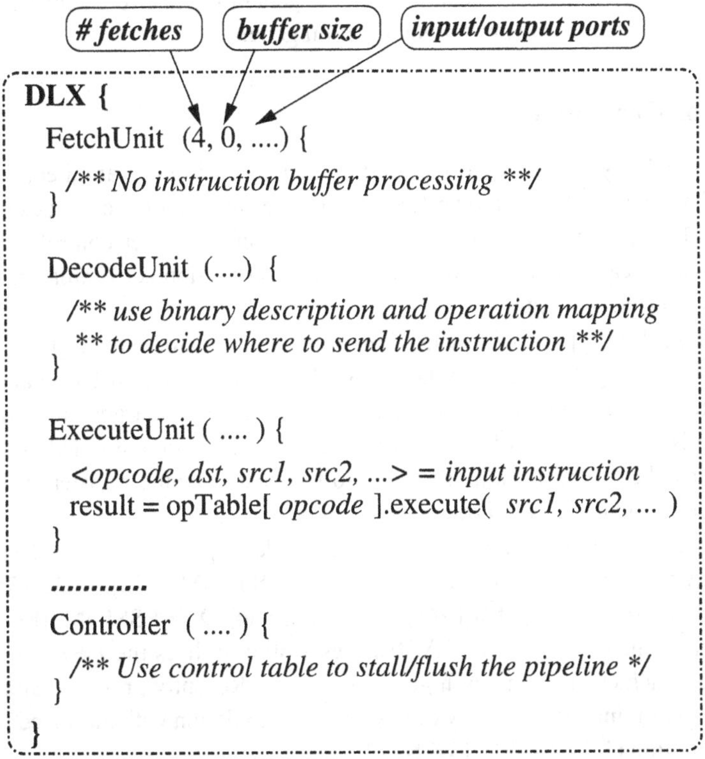

 # fetches **buffer size** **input/output ports**

```
DLX {
  FetchUnit (4, 0, ....) {
    /** No instruction buffer processing **/
  }

  DecodeUnit (....) {
    /** use binary description and operation mapping
     ** to decide where to send the instruction **/
  }

  ExecuteUnit ( .... ) {
    <opcode, dst, src1, src2, ...> = input instruction
    result = opTable[ opcode ].execute( src1, src2, ... )
  }

  ...........

  Controller ( .... ) {
    /** Use control table to stall/flush the pipeline */
  }
}
```

Figure 4.8: Simulation model generation for the DLX architecture

The generated simulation models combined with the existing simulation kernel creates a cycle-accurate structural simulator that executes the assembly instructions. In our framework, the assembly instructions generated by the EXPRESS compiler [4] are loaded into the instruction memory of the simulator.

We have used the generated reference models in two top-down validation scenarios. Chapter 5 describes design validation using equivalence checking between the implementation and the generated hardware model. Chapter 6 presents test generation and functional validation using the generated simulation models. We have also used the generated simulation and hardware models for design space exploration (DSE) of programmable architectures. We briefly outline the exploration methodology in Appendix E.

4.4 Related Work

We discuss related work on executable model generation in two categories: simulator generation and hardware implementation generation.

Simulator Generation

An extensive body of recent work has addressed instruction-set architecture simulation. The wide spectrum of today's instruction-set simulation techniques includes the most flexible but slowest interpretive simulation and faster compiled simulation. Recent research addresses retargetability of instruction-set simulators using a machine description language.

Simplescalar [51] is a widely used interpretive simulator. Shade [113], Embra [25] and FastSim [23] based simulators use dynamic binary translation and result caching to improve simulation performance. A fast and retargetable simulation technique is presented in [65]. It improves traditional static compiled simulation by mapping the target machine registers to the host machine registers through a low level code generation interface.

Retargetable simulators based on processor description languages have been proposed within the framework of FACILE [24], Sim-nML [72], ISDL [31], MIMOLA [117], ANSI C [27], LISA ([9], [121]), and EXPRESSION ([8], [102]). The simulator generated from a FACILE description utilizes the *Fast Forwarding* technique to achieve reasonably high performance. Recently proposed simulation techniques combine the flexibility of interpretive simulation with the speed of compiled simulation (JIT-CCS [9], IS-CS [77]).

Hardware Generation

There are two major approaches in the literature for synthesizable HDL generation. The first one is a parameterized processor core based approach. These cores are bound to a single processor template whose architecture and tools can be modified to a certain degree. The second approach is based on processor specification languages.

Examples of processor template based approaches are Xtensa [130], Jazz [46], and PEAS [75]. Xtensa [130] is a scalable RISC processor core. Configuration options include the width of the register set, caches, memories etc. New functional units and instructions can be added using the Tensilica Instruction Language (TIE). A synthesizable hardware model along with software toolkit can be generated for this class of architectures. Improv's Jazz [46] processor is a VLIW processor that permits the modeling and simulation of a system consisting of multiple processors, memories, and peripherals. It allows modifications of data width, number of registers, depth of hardware task queue, and addition of custom functionality in Verilog. PEAS [75] is a GUI based hardware/software codesign framework. It generates HDL code along with software toolkit. It has support for several architecture types and a library of configurable resources.

Processor description language driven HDL generation frameworks can be divided into three categories based on the type of information the languages can capture. Structure-centric ADLs such as MIMOLA are suitable for hardware generation. Some of the behavioral languages (such as ISDL and nML) are also used for hardware generation. For example, the HDL generator HGEN [30] uses ISDL description, and the synthesis tool GO [49] is based on nML. Itoh et al. [76] have proposed a micro-operation description based synthesizable HDL generation.

Mixed languages such as LISA and EXPRESSION capture both structure and behavior of the processor. The synthesizable HDL generation approach based on LISA language [84] produces an HDL model of the architecture. The designer has the choice to generate a VHDL, Verilog or SystemC representation of the target architecture [85]. The HDL generation technique presented in this chapter combines the advantages of the processor template based environments and the language based specifications using the EXPRESSION ADL.

4.5 Chapter Summary

A major challenge in a top-down validation methodology is the development of a functional abstraction technique that is able to generate executable models from the specification for a wide variety of programmable architectures including RISC, DSP, VLIW, and superscalar. We have studied the similarities and differences of each architectural feature in different architecture domains. Based on our observations we have defined generic functions, sub-functions, and computational environment needed to capture a wide variety of programmable architectures.

Our functional abstraction technique enables model generation for simulation, hardware generation, and property checking from the ADL specification. The generated models are used for design validation, test generation and design space exploration of programmable architectures.

5

DESIGN VALIDATION

One of the major challenges in validation of programmable architectures is the verification of RTL design (implementation). Design validation techniques can be broadly categorized into simulation-based approaches and formal techniques. Due to the complexity of modern designs, validation using only traditional scalar simulation cannot be exhaustive. Formal techniques exhaustively analyze parts of the design but, because of state space explosion, are not suitable for the complete design. *Equivalence Checking* is one of the most widely used formal techniques in industry today. Typically, the implementation is compared with a set of Boolean equations, or an optimized circuit is compared with the original circuit. *Symbolic simulation* has proven to be an efficient technique, bridging the gap between traditional simulation and full-fledged formal verification.

Figure 1.6 shows a traditional architecture validation flow. The implementation design is validated using a combination of simulation techniques and formal methods. The existing techniques employ a bottom-up approach to validation, where the functionality of an existing processor is, in essence, reverse-engineered from its RTL implementation. The validation technique presented in this chapter is complementary to these bottom-up approaches.

Figure 5.1 shows our top-down validation methodology. Our validation framework allows generation of synthesizable RTL description as well as specific properties. The RTL description can be used for checking equivalence with the given implementation. However, generation of specific behaviors would enable property checking. For example, our framework generates the property: $output = \sum_{i=1}^{n} input_i$, for a n-input adder. The design should satisfy this property irrespective of the adder implementation, such as ripple-carry adder or carry look-ahead adder.

A major advantage of property checking is that it reduces the complexity of verification. However, this technique raises an important question: how to choose the set of properties. A set of properties can be chosen in two different ways. First, the designers can decide what properties are important to be verified for the design based on their design knowledge and past experience. They can then choose the properties to uncover otherwise difficult-to-find bugs. Second, a set of behaviors can be chosen and their effectiveness can be evaluated. For example, to verify a memory controller in a microprocessor, it is necessary to generate properties to validate each output of the controller. To measure the effectiveness of these properties, a set of coverage measures can be used during property checking [36].

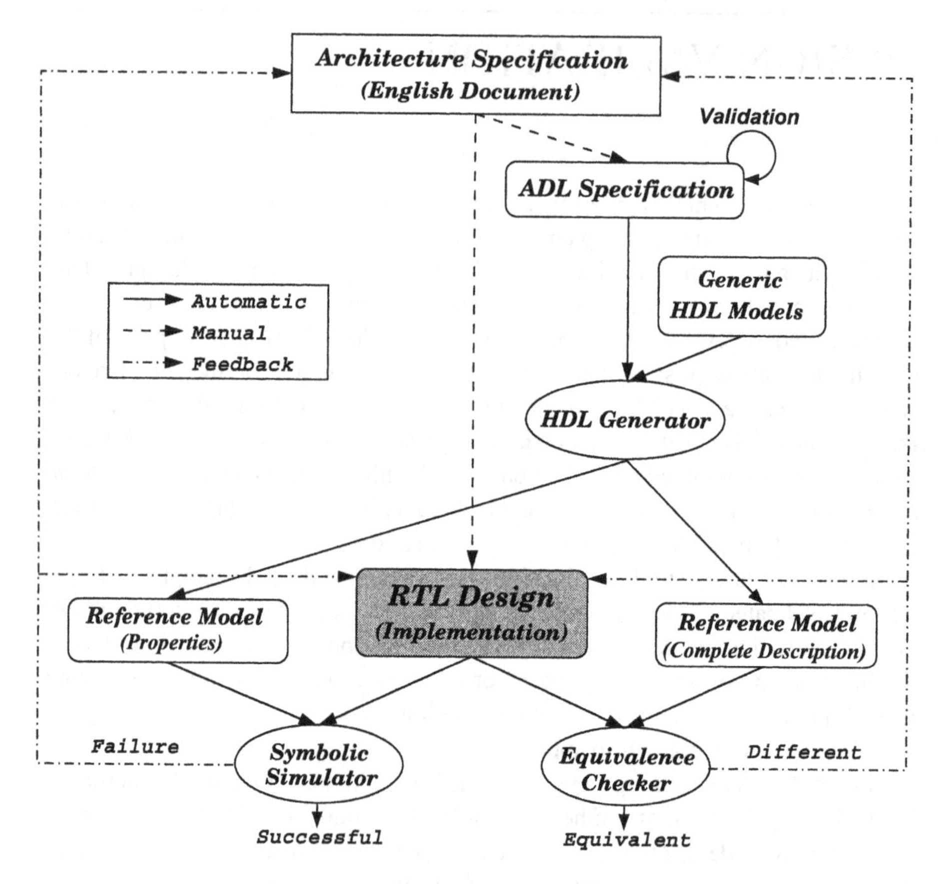

Figure 5.1: Top-down validation methodology

This chapter is organized as follows. Section 5.1 describes property checking using symbolic simulation. Section 5.2 describes validation using equivalence checking. Section 5.3 presents the validation experiments followed by the related work in Section 5.4. Finally, Section 5.5 summarizes the chapter.

5.1 Property Checking using Symbolic Simulation

Symbolic simulation combines traditional simulation with formal symbolic manipulation [111]. Each symbolic value represents a signal value for different operating conditions, parameterized in terms of a set of symbolic Boolean variables. By this encoding, a single symbolic simulation run can cover many conditions that would require multiple runs of a traditional simulator.

Figure 5.2(a) shows a simple n-input *AND* gate. Exhaustive simulation of the *AND* gate requires 2^n binary test vectors. However, the ternary simulation uses (*0, 1, x*) and requires $(n+1)$ test vectors for the *AND* gate. Figure 5.2(b) shows the vectors: n vectors with one input set to '0' and the remaining inputs set to 'x', and one vector with all inputs set to '1'. Finally, symbolic simulation [111] requires only one vector using n symbols ($s_1, s_2, ..., s_n$) as shown in Figure 5.2(c).

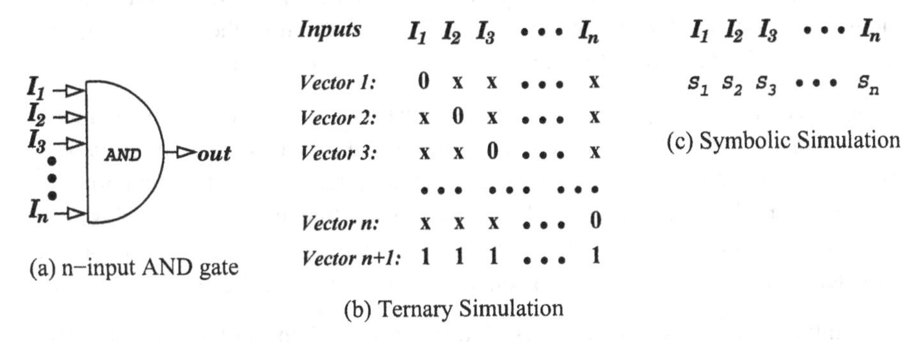

Figure 5.2: Test vectors for validation of an *AND* gate

Researchers at IBM first introduced symbolic simulation to reason about properties of circuits described at the register-transfer level. With the advent of Binary Decision Diagrams (BDDs), the technique became much more practical. Providing a canonical representation for Boolean functions, BDDs enabled the implementation of an efficient event-driven logic simulator that operated over a symbolic domain. By encoding a model's finite domain using a Boolean encoding, it is possible to symbolically simulate the model using BDDs. Bryant's formal state transition model for a ternary system [112], and Seger's work on symbolic trajectory evaluation renewed further interest in symbolic execution [14].

The symbolic simulator (used in our framework) uses symbolic trajectory evaluation (STE). In this section we informally describe STE. The formal description of STE is available in [14]. STE is a modified form of symbolic simulation that operates over the quaternary logic domain 0, 1, X, and T [14]. A state of the circuit is defined as the set of all node values at a particular time instant. The value domain is partially ordered and forms a complete lattice, $X \sqsubseteq 0$ indicates X has less information than 0, or X is weaker than 0. The information content of 0 and 1 are not comparable. If $r \sqsubseteq q$ and $r \sqsubseteq t$, we can think of r as representing both q and t. Any property that holds for a state such as r will also hold for all the states above it in the lattice, for example q and t.

STE provides a mathematically rigorous method for establishing that properties (assertions) of the form *antecedent (A)* $\Rightarrow$ *consequent (C)* hold for a given simulation model of a circuit. For the test vector shown in Figure 5.2(c), the antecedent is: (I_1 is s_1, I_2 is s_2, ..., I_n is s_n) from time 0 to 1, and the consequent is: *out* is $s_1 \& s_2 \& ... \& s_n$ from time 1 to 2. Circuit state holders are initialized with symbolic values specified by the antecedent. The model is then simulated, typically for one or two clock cycles, while driving the inputs with symbolic values during simulation. The resulting values, appear on selected internal nodes and primary outputs, are compared with the expected values expressed in the consequent. In general, the values could be functions over a finite set of variables.

A trajectory is a sequence of states such that each state has at least as much information as the next-state function applied to the previous state. Intuitively, a trajectory is a state sequence constrained by the system's next-state function. A successful simulation of assertion $A \Rightarrow C$ establishes that any sequence of assignments of values to circuit nodes that is both consistent with the circuit behavior and consistent with antecedent A is also consistent with consequent C.

Symbolic trajectory evaluation is used to verify that an implementation satisfies its specification (reference model). Necessary assertions are extracted from the reference model. If the implementation (e.g., *RTL design*) is correct, these assertions should hold during symbolic simulation of the *RTL design*. An assertion ($A \Rightarrow C$) holds if the weakest antecedent trajectory that the implementation goes through during simulation (using A) is at least as strong as the weakest sequence satisfying the consequent C. Informally, the outputs produced during simulation (using A) should be at least as strong as the expected outputs (given in C).

To verify that the implementation satisfies certain properties, our framework generates the intended properties instead of generating the complete reference design. We use Versys2 [82] that uses symbolic trajectory evaluation to perform property checking. The assertions are automatically generated from the reference model [69]. Versys2 symbolically simulates the implementation by using the generated assertions to ensure that the implementation satisfies the reference model. A counter-example is generated if an assertion fails in the implementation. The feedback is used to modify the implementation.

5.2 Equivalence Checking

Equivalence Checking is a branch of static verification that employs formal techniques to prove that two versions of a design either are, or are not, functionally equivalent. The equivalence checking flow consists of four stages: *reading, matching, verification* and *debugging.* The matching and verification stages are those most impacted by design transformations. During the *reading* stage, both versions of the design are read by the equivalence checking tool and segmented into manageable sections called logic cones. Logic cones are groups of logic bordered by registers, ports, or black boxes. Figure 5.3(a) shows the cones for a typical design block. The output border of a logic cone is referred to as the compare point. For example, OUT_1 is the compare point in $Cone_1$ of Figure 5.3(a).

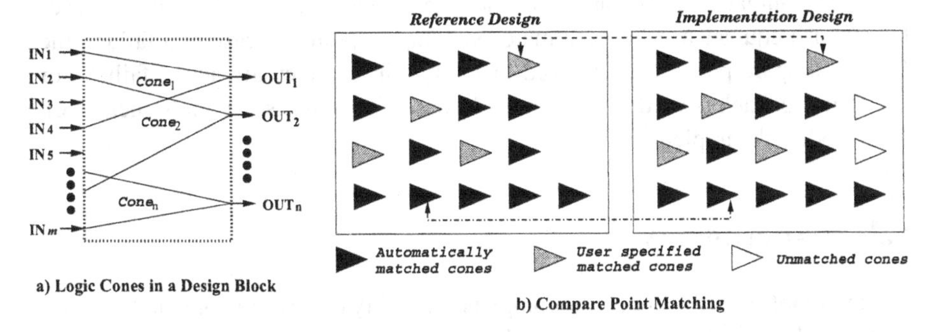

Figure 5.3: Compare point matching between reference and implementation design

In the *matching* phase, the tool attempts to match, or map, compare points from the reference design to their corresponding compare point within the implementation design [18]. Two types of matching techniques are used: name based (non-function) and function based (signature analysis). Figure 5.3(b) shows compare point matching for a typical reference design and implementation. For better performance, the majority of the matching should be completed by more efficient name based methods.

Design transformations can result in fewer cones being matched by the name based techniques, slowing match performance. Creating compare rules assist name based techniques, but determination and creation of the rules themselves can be time consuming. If the implementation is drastically different than the reference design, design rules cannot be written and compare points have to be manually matched for better performance or matched using more costly function based techniques. This becomes impractical for design with many unmatched points.

During the *verification* stage, each matched compare point is proven either functionally equivalent or non-equivalent ([12], [58]). Design transformations can impact the structure of a logic cone in the implementation design. When logic cones are very dissimilar, performance suffers. In some cases, such as during retiming, the logic cones can change so significantly that additional setup is required to successfully verify the designs. The *debugging* phase begins when the tool has returned a non-equivalent result. Design transformations that have not been accounted for can lead to a false negative result, and valuable time could be spent debugging designs that are, in reality, equivalent. The solution would be to perform additional setup so that the tool is guided for the given designs.

Our framework generates the synthesizable RTL description to enable equivalence checking using Synopsys Formality [128]. The tool reads both the reference and the implementation designs, and attempts to match the compare points between them. The unmatched compare points need to be mapped manually. The tool tries to establish equivalence for each matched compare point. In case of a failure, the failing compare points are analyzed to verify whether they are actual failures or not. The feedback is used to perform additional setup (in case of a false negative), or to modify the implementation.

5.3 Experiments

An important aspect of our methodology is the ability to perform both model (property) checking and equivalence checking depending on the generated reference model. Our validation framework uses the Versys2 [82] symbolic simulator and the Formality [128] equivalence checker. Section 5.3.1 presents validation of a memory management unit of a microprocessor that is compliant with the PowerPC instruction-set using symbolic simulation. Section 5.3.2 presents the validation of a RISC DLX processor using equivalence checking.

5.3.1 Property Checking of a Memory Management Unit

The memory management unit (MMU) typically supports demand-paged virtual memory. It consists of blocks such as *Segment Registers*, *Translation Lookaside Buffers (TLBs)*, and *Block Address Translation (BAT)* arrays. Each of these memory blocks are composed of sub-blocks. For example, a TLB has three sub-blocks: *entry* (data information), *LRU* (least recently used information), and *valid* (information regarding validity of the data) as shown in Figure 5.4. Each of these sub-blocks is implemented as SRAM. The typical operations in SRAM are read and write. Therefore, a natural property to verify is to check read and write for each

SRAM cell. The generated reference model contains the following Verilog code segment to verify the read and write properties for an SRAM cell.

```
// Write Property
always @ (wrClk or wrEn or dIn or wrAddr)
begin
  if (wrClk & wrEn) ram[wrAddr] <= dIn;
end


// Read Property
assign out = (rdClk & rdEn) ? ram[rdAddr] : 32'b0;
```

The Versys2 symbolic simulator does not have automatic node matching (compare point matching) scheme. Therefore, it is necessary to manually map the nodes between the reference model and the implementation. We modified the Versys2 configuration file to provide the node mapping between the reference model and the implementation. For example, the *wrClk* of the reference model is mapped to *sramWrClk* of the implementation. An interesting feature of this validation approach is that the same set of properties (without any modification) is applied to all MMU memory blocks. However, in each case, the node mapping must be modified.

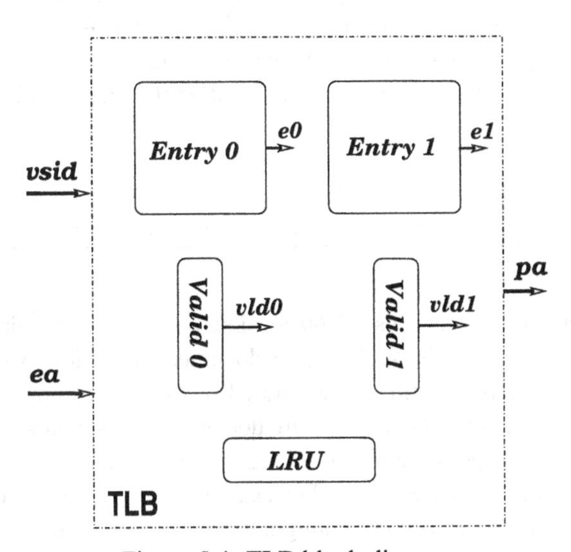

Figure 5.4: TLB block diagram

To verify whether the *RTL design* correctly implements the TLB miss detection, our framework generated the following Verilog code segment. The information needed to build this property is directly available from the specification of the MMU.

```
assign inp =({1'b1,vsid[0:23],ea[4:9],ea[10:13]});
assign out0=({vld0,e0[0:23],e0[24:29],e0[54:57]});
assign out1=({vld1,e1[0:23],e1[24:29],e1[54:57]});
assign hit0=(inp == out0);
assign hit1=(inp == out1);
assign miss=~(hit0 | hit1);
```

This property verifies miss detection for a two-way set-associative TLB. It would be a simple extension for generating this property for a n-way set-associative TLB. Here *vsid* (virtual segment id) and *ea* (effective address) are inputs, and *pa* (physical address) is the output of the TLB block. The *e* and *vld* variables are outputs from the *entry* and *valid* blocks respectively as shown in Figure 5.4.

Similarly we have generated and validated the property for the BAT array miss detection. There are several mismatches found (between the reference model and the implementation) during property checking. For instance, the architecture specification document does not provide the value for the else condition (default value of a signal for example) in most of the cases. As a result the description of the property does not have the default value for a signal, whereas the signal has a definite value in its implementation under all possible conditions. Symbolic simulation produced mismatches in those cases. Consider the following read implementation of a SRAM cell.

```
assign out = (rdClk & rdEn) ? ram[rdAddr] : 32'b0;
```

This implementation assigns *32'b0* to signal *out* when condition (*rdClk & rdEn*) is *false*. However, the architecture document does not specify the value in the default case. As a result, the generated property does not have this value that caused the mismatch. The architecture document is updated to add the values in all cases. It is also possible to impose certain constraints in Versys2 to avoid the detection of such false negatives. For example, we can set the condition (*rdClk & rdEn*) as *true* in the Versys2 configuration file to avoid the detection of the mismatch mentioned above.

5.3.2 Equivalence Checking of the DLX Architecture

We validated the DLX [55] processor using equivalence checking. We obtained a VHDL description of the synthesizable 32-bit RISC DLX from *eda.org* [44] and used it as the *implementation*. Our framework generated the VHDL description from the ADL specification using the method described in Section 4.3. The generated VHDL description is used as the *reference model* (specification) for the validation.

Regardless of the implementation style, the equivalence checker can verify a design based on the correct behavior in the reference model. For example, our HDL generation framework generates a 32-bit carry look-ahead (CLA) adder. The equivalence checker verifies that this design is equivalent to the 32-bit adder implementation, which uses a ripple-carry adder principle. Equivalence checking took 4 seconds to verify the adder on a 300 MHz Sun Ultra-250 with 1024M RAM. Similarly, we generated a structural model of a 32×32 register file and used it as a reference model to verify the behavioral register file implementation [44]. In this case, equivalence checking took 432 seconds. The majority of this time (347 seconds) was consumed in the elaboration (linking) phase of the behavioral implementation.

Our framework generated synthesizable RTL for 32-bit RISC DLX that supports signed operations. To avoid memory explosion during equivalence checking, we guided the RTL generation process to have a structure similar to the implementation [44]. The equivalence checking process took 397 seconds. We have encountered a mismatch in the output data bus at clock cycle 2500. The analysis revealed that the problem is in the overflow bit of the adder. The ripple-carry adder implementation of the DLX [44] had an incorrect computation of the overflow bit.

Design analysis in our framework is easy once we figure out the module that is causing the problem. For example, in this particular case once we know that the adder is causing the problem, we can verify the adder implementation of the DLX by generating an adder specification (HDL description) from our framework and applying equivalence checking.

Table 5.1: Validation of the DLX implementation using equivalence checking

Reference **Implementation**	32-bit CLA adder ripple-carry adder	32×32 register-file behavioral model	32-bit DLX DLX [44]
Validation Time	4 seconds	432 seconds	397 seconds

Table 5.1 summarizes the experimental results. Each column in the table presents the equivalence checking time for the respective reference model and the imple-

mentation. As we can see from the table that the validation time is longer for equivalence checking of the register file than the DLX processor. This is due to the fact that the models used for verifying the register-file are very different (structural vs. behavioral). However, we have guided the reference model generation process of the DLX processor such that the reference model has structure similar to that of the implementation.

5.4 Related Work

Several approaches for formal or semi-formal verification of programmable architectures have been developed in the past. Theorem proving techniques, for example, have been successfully adapted to verify pipelined processors ([20], [62], [78]). However, these approaches require a great deal of user intervention, especially for verifying control intensive designs. Hosabettu [118] proposed an approach to decompose and incrementally build the proof of correctness of pipelined microprocessors by constructing the abstraction function using completion functions.

Burch and Dill presented a technique for formally verifying pipelined processor control circuitry [53]. Their technique verifies the correctness of the implementation model of a pipelined processor against its instruction-set architecture (ISA) model based on quantifier-free logic of equality with uninterpreted functions. The technique has been extended to handle more complex pipelined architectures by several researchers ([64], [79]). The approach of Velev and Bryant [79] focuses on efficiently checking the commutative condition for complex microarchitectures by reducing the problem to checking equivalence of two terms in a logic with equality, and uninterpreted function symbols.

Huggins and Campenhout verified the ARM2 pipelined processor using abstract state machine [56]. Levitt and Olukotun [57] presented a verification technique, called unpipelining, which repeatedly merges last two pipeline stages into one single stage, resulting in a sequential version of the processor. A framework for microprocessor correctness statements about safety that is independent of implementation representation and verification approach is presented in [71].

Ho et al. [90] extract controlled token nets from a logic design to perform efficient model checking. Jacobi [13] used a methodology to verify out-of-order pipelines by combining model checking for the verification of the pipeline control, and theorem proving for the verification of the pipeline functionality. Compositional model checking is used to verify a processor microarchitecture containing most of the features of a modern microprocessor [115]. There has been a lot a work in the area of module level validation such as verification of floating-point unit [91], and protocol validation such as verification of cache coherence protocol [29].

5.5 Chapter Summary

Functional verification is one of the most complex and expensive tasks in the current microprocessor design flow. A significant bottleneck in the validation of such systems is the lack of a golden reference model. Thus, many existing approaches employ a bottom-up validation methodology by using a combination of simulation techniques and formal methods.

This chapter presented a top-down validation methodology driven by an ADL. The reference model (HDL description) is generated from the ADL specification. An important aspect of our methodology is the ability to perform both model (property) checking and equivalence checking depending on the generated reference model. Our framework generates the intended properties to enable model checking, and generates the RTL description of the processor to enable equivalence checking. To verify the properties, the framework uses Versys2 [82] that generates assertions from the reference model and applies them to the implementation using symbolic trajectory evaluation. The framework uses Formality [128] to perform equivalence checking. We have applied our methodology in two validation scenarios: property checking of a memory management unit of a microprocessor that is compliant with the PowerPC instruction-set, and equivalence checking of the DLX architecture. We have identified a functional bug in the ripple-carry adder module of the DLX implementation [44].

Specification-driven hardware generation and validation of design implementation using equivalence checking has one limitation: the structure of the generated hardware model (reference) needs to be similar to that of the implementation. This requirement is primarily due the limitation of the equivalence checkers available today. Equivalence checking is not possible using these tools if the reference and implementation designs are large and drastically different. Property checking can be useful in such scenarios to ensure that both designs satisfy a set of properties. However, property checking does not guarantee equivalence between two designs. As a result, it is also necessary to use other complementary validation techniques (such as simulation) to verify the implementation.

6

FUNCTIONAL TEST GENERATION

As embedded systems continue to face increasingly higher performance requirements, deeply pipelined processor architectures are being employed to meet desired system performance. Functional validation of such programmable processors is widely acknowledged as a major bottleneck in current design methodology. Simulation is the most widely used form of microprocessor verification: millions of cycles are spent during simulation using a combination of random and directed test cases in traditional design flow. Certain heuristics and design abstractions are used to generate directed random testcases. However, due to the bottom-up nature and localized view of these heuristics the generated testcases may not yield a good coverage. The problem is further aggravated due to the lack of a comprehensive functional coverage metric.

This chapter presents two specification-driven test generation techniques. Section 6.1 describes a model checking based functional test program generation technique for pipelined processors. Section 6.2 proposes a functional fault model that is used to define the functional coverage of pipelined architectures. The fault model is used for coverage-driven test generation. Section 6.3 presents related work addressing validation of pipelined processors. Finally, Section 6.4 summarizes the chapter.

6.1 Test Generation using Model Checking

This section presents a specification-driven test generation technique for pipelined processors. To make ADL-driven test generation applicable to realistic embedded processors, three important steps must be automated using efficient techniques. First, the processor model generation from the specification needs to be automated. Second, there is a need for a comprehensive functional coverage metric that can be used to generate test programs. Finally, an efficient test generation technique is necessary that can model complex designs and generate functional test programs.

6.1.1 Test Generation Methodology

Figure 6.1 shows our graph based functional test program generation methodology. The graph model of the processor is generated from the ADL specification. The properties are generated based on the graph coverage metric discussed later in this section. The properties are applied at the module level using the SMV model checker [43]. The counterexamples are analyzed to generate test programs at the processor level. We apply these test programs to the simulator of the processor to ensure that the coverage criteria is met. If necessary, additional properties can be added manually. This technique reduces the time and space required for generating test programs by applying properties at the module level and composing the responses in sequence by traversing the pipeline graph.

```
Algorithm 7: Test Program Generation
Inputs: ADL specification of the pipelined processor
Outputs: Test programs to verify the pipeline behavior.
Begin
   Generate graph model of the architecture.
   Generate properties based on the graph coverage
   for each property prop for graph node n
       inputs = ϕ
       while (inputs != primary_inputs)
              Apply prop on node n using SMV model checker
              inputs = Find i/p requirements for n from counterexample
              if inputs are not primary_inputs
                 Extract output requirements for parent of node n
                 prop = modify prop with new output requirements
                 n = parent of node n
              endif
       endwhile
       Convert primary input assignments to a test program
       Generate the expected output using a simulator.
   endfor
   return the test programs
End
```

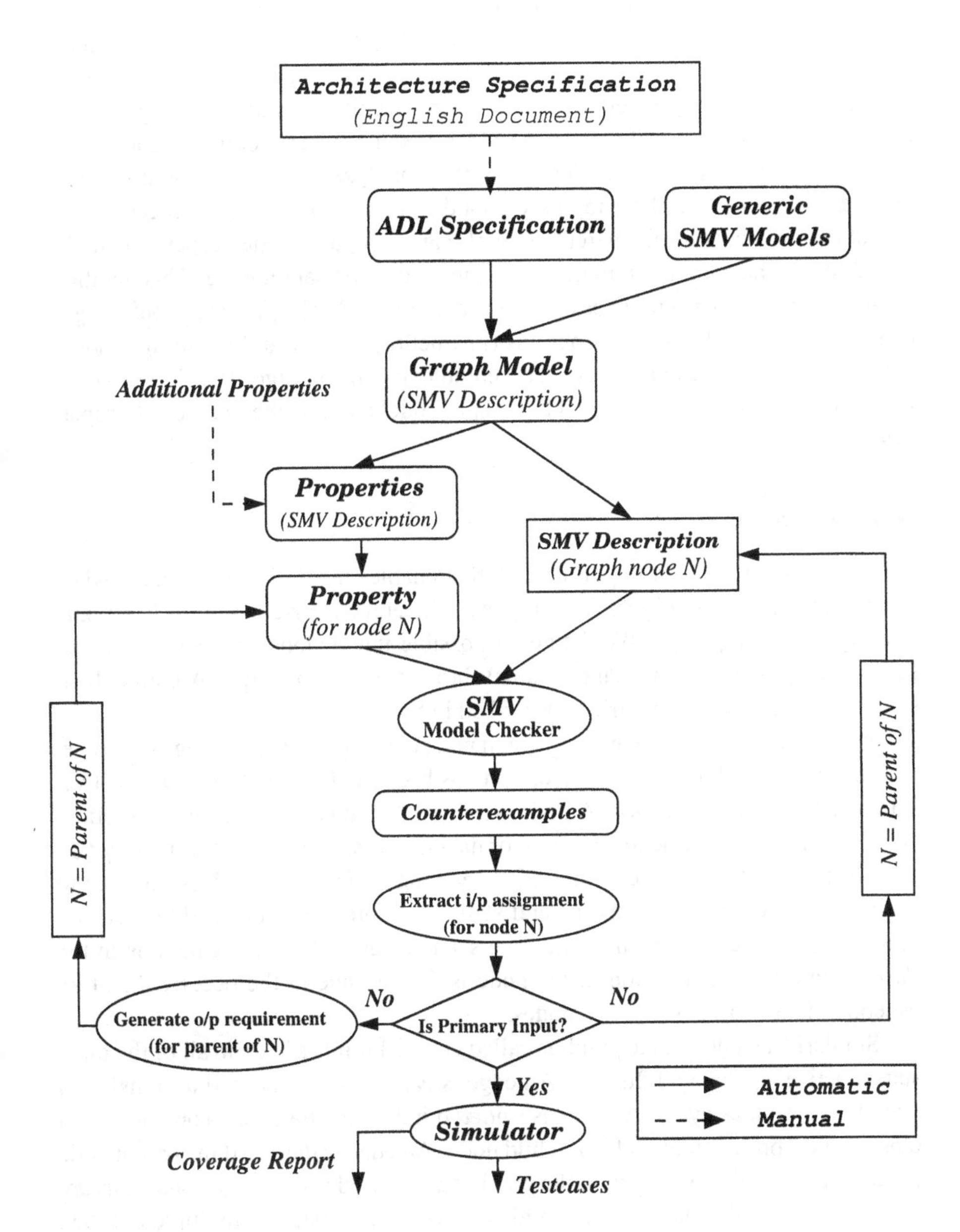

Figure 6.1: Test program generation methodology

Algorithm 7 presents our specification driven test generation procedure. A property *prop* is applied to a module corresponding to node n in the graph model. The framework actually generates the negation of the properties that we want to verify. For example, to generate a testcase for assigning a value *5* to a register *R7*, the property states that "*R7 != 5*". The SMV model checker produces a counterexample for the property. The counterexample is analyzed to find the input requirements for the node n. If these inputs are not the primary inputs of the processor, the output requirements for the parent node of n are computed. The property is modified based on the output requirements and applied to the parent node. This iteration continues until primary input assignments are obtained. The primary input assignments are converted into test programs (instruction sequences) by putting random values in the unassigned inputs. The complexity of the algorithm is $O(n \times p)$, where n is the number of nodes in the graph model and p is the number of properties.

Graph Coverage

Measuring progress is an important task that enables the designer to decide when to end the verification effort. We propose a coverage metric based on functional coverage of the pipeline. We define all possible interactions between operations (instructions) and pipeline stages (paths) through graph coverage. A comprehensive functional coverage metric is described in Section 6.2.

We define graph coverage using graph node coverage and graph edge coverage. A node in the graph is called covered if it has been in all of the four states: active, stalled, exception and flushed. A node is *active* when it is executing an instruction. A node can be *stalled* due to structural or data hazards. A node can be in *exception* state if it generates an exception while executing an instruction. It is possible to have multiple exception scenarios and stall conditions for a node. However, our current node coverage requires only one scenario in each case. A node is in the *flushed* state if an instruction in the node is flushed due to the occurrence of an exception in any of its successor nodes.

Similarly, an edge in the graph is called covered if it has been in all of the three states: active, stalled and flushed. An edge is *active* when it is used to transfer an operation in a clock cycle. An edge is *stalled* if it does not transfer an operation in a clock cycle from a parent node to a child node. An edge is *flushed* if the parent node is flushed due to the exception in the child node. The edge coverage conditions are redundant if a node has only one child. However, if a node has multiple children (or parents), edge coverage conditions are necessary.

Our test generation algorithm traverses the pipeline graph and generates properties based on the graph coverage described above. For example, consider the test generation for a feedback path (edge) from *MUL7* to *IALU* for the DLX architecture shown in Figure 3.6. To generate a test for making the feedback path *active*, two properties are generated: i) make the node *MUL7* active in clock cycle *t*, and ii) make the node *IALU* active in clock cycle *(t+1)*. This would lead to a test program that has a multiply operation followed by six NOPs (no operation), and finally an add operation.

6.1.2 A Case Study

In a case study we successfully applied the proposed methodology to the DLX processor [55]. Figure 3.6 shows the graph model of the DLX processor. First, we present the test program generation results for the DLX processor. Next, we describe a test generation scenario using an illustrative example to demonstrate the efficiency of our technique.

Test Generation Results

This section describes the number of test cases generated for the DLX processor using the graph coverage described in Section 6.1.1. The DLX processor shown in Figure 3.6 has 20 nodes and 24 edges (except feedback paths). We have described all the 91 instructions of the DLX processor [55].

Table 6.1: Number of test programs in different categories

Node Coverage				Edge Coverage		
Active	Stalled	Flushed	Exception	Active	Stalled	Flushed
91	20	20	20	24	24	24

Table 6.1 shows the number of test programs generated for the node and edge coverage of the DLX processor. Although, 20 testcases would suffice for the *active* node coverage, we use 91 test cases in this category to cover all the instructions. Also, there are many ways of making a node stalled, flushed or in exception condition. We chose one such scenario. If we consider all possible scenarios, the number of test programs will increase. In this case, our algorithm generated 223 test programs in 91 seconds on a 333 MHz Sun UltraSPARC-II with 128M RAM.

As mentioned earlier, some of the test programs are redundant. For example, since there are four pipeline paths, we need only four test programs that exercise the four paths. These four test programs will make all the nodes *active*. Similarly, if we assume VLIW DLX, the decode node will be stalled if any one of its four children is stalled. Furthermore, if the MEM node is stalled, all of its four parents will also be stalled. This implies that we need only 14 testcases for node stalling. Likewise, if the MEM node is in exception, the instructions in all the predecessor nodes will be flushed. Hence, we need only 2 testcases for flushing. Finally, some of the node coverage testcases also satisfies the edge coverage. We need a total of 43 test programs in this case. Table 6.2 shows the number of reduced test programs in different categories.

Table 6.2: Reduced number of test programs

Node Coverage				Edge Coverage		
Active	Stalled	Flushed	Exception	Active	Stalled	Flushed
4	14	2	20	4†	14† + 3	2†

Test Program Generation: An Example

Example 6.1: *Consider a fragment of the DLX pipeline containing three internal registers of the division unit (DIV) as shown in Figure 6.2. The goal is to initialize two registers* A_{in} *and* B_{in} *with values 2 and 3 respectively at clock cycle 9.*

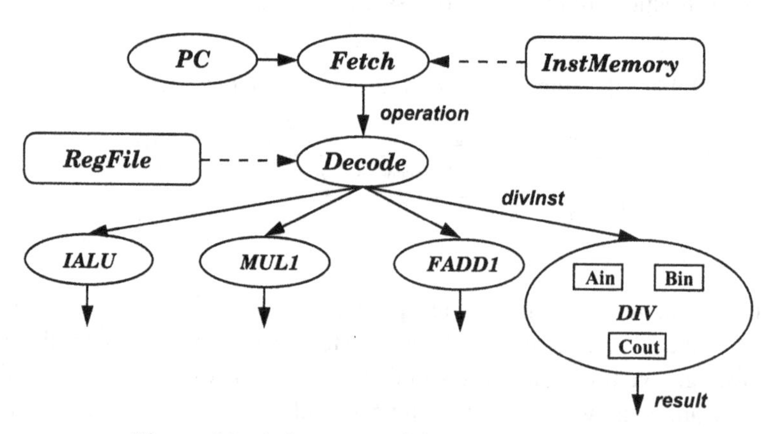

Figure 6.2: A fragment of the DLX architecture

In this section we describe our test generation approach using Example 6.1. The two internal input registers for DIV unit are A_{in} and B_{in}. The internal output register for DIV unit is C_{out}. The input instruction is *divInst* and the output is

† Same testcases as in the node coverage.

result. In this particular scenario, A_{in} and B_{in} receive data from the first and second source operands of the input instruction (*divInst*) i.e., $A_{in} = divInst.src1$ and $B_{in} = divInst.src2$; C_{out} returns the result of the division i.e., $C_{out} = A_{in} \div B_{in}$; finally the output is generated from C_{out} i.e., $result = C_{out}$.

The following property generates the instruction sequence to initialize A_{in} and B_{in} with values 2 and 3 respectively at clock cycle 9. The property is written using the SMV language [43]. Informally speaking, it implies that if the current clock cycle is 8, in the next cycle $DIV.A_{in}$ should not be 2 or $DIV.B_{in}$ should not be 3:

```
assert G((cycle = 8) -> X((DIV.Ain ~= 2) | (DIV.Bin ~= 3)));
```

If this property is applied to the complete description of the processor, SMV takes 375.98 seconds on a 333 MHz Sun UltraSPARC-II with 128M RAM, and requires 1928568 BDD nodes to generate the counterexample. In the remainder of this section, we illustrate how our test generation methodology improves both time and space requirements for the Example 6.1.

We modify this global property to make it applicable at module level (as shown below) and apply the property to the division unit (*DIV*) using SMV:

```
assert G((cycle=8) -> X((Ain ~= 2) | (Bin ~= 3)));
```

The next step is to analyze the counterexample produced by SMV to extract the input requirements for the division unit. For example, in this case the input requirements are simple: *divInst.src1 = 2* and *divInst.src2 = 3*. These input requirements are used to generate the expected output assignments for the decode unit (parent of the division unit). Also, the cycle count requirement is modified for the decode unit. The modified property (shown below) is applied to the decode unit.

```
assert G((cycle=7) -> X((divInst.src1 ~= 2) |
                        (divInst.src2 ~= 3)));
```

The counterexample is analyzed to extract the input requirements for the decode unit. The decode has two inputs: *operation* and *RegFile*. For example, in this case the input requirements are: *operation.opcode = DIV, operation.src1 = 1, operation.src2 = 2, RegFile[1] = 2*, and *RegFile[2]=3*. This indicates that the *operation* should be a division operation with *src1* as R1 and *src2* as R2. It also implies that the register file should have the values 2 and 3 at locations 1 and 2 respectively.

There are two tasks to be done here. First, initialize a register file location with a specific value at a given clock cycle *t*. It is done using a *move-immediate* instruction fetched at *(t-5)*. In this case, the *move-immediate* operations should be done at clock cycle 2 and 3 to make the data available at cycle 8. The second task is to convert the remaining input requirements as the expected outputs for the fetch unit (parent of the decode). The modified property (shown below) is applied to the fetch unit.

```
assert G((cycle=6) -> X((operation.opcode ~= DIV) |
                         (operation.src1 ~= 1)    |
                         (operation.src2 ~= 2)));
```

The counterexample is analyzed to extract the input requirements for the fetch unit. The fetch unit has two inputs: *PC* and instruction memory. The expected value for PC is 5 and InstMemory[5] has instruction: *DIV* R_x R_1 R_2. These are primary inputs of the processor. The final test program, shown below, is constructed by putting random values in the unspecified fields:

```
Fetch Cycle     Opcode Dest Src1 Src2        Comments
-----------     ------ ---- ---- ----     --------------
 1               NOP                      R0 is always 0
 2               ADDI  R1,  R0,  #2       R1 = 2
 3               ADDI  R2,  R0,  #3       R2 = 3
 4               NOP
 5               NOP
 6               NOP
 7               DIV   R3,  R1,  R2
```

For this example, the system took less than a second to come up with the counterexample on a 333 MHz Sun UltraSPARC-II with 128M RAM. This time includes the time taken by SMV in verifying three module level properties. It also includes the time taken by our system in traversing the graph and generating the new properties with input/output computations using counterexamples. The total number of BDD nodes allocated is 5600. If the property is applied to the complete description of the processor, SMV takes 375.98 seconds and requires 1928568 BDD nodes to generate the counterexample. Clearly, our technique reduced the test generation time and the required BDD size by an order of magnitude.

6.2 Functional Coverage driven Test Generation

Several coverage measures are commonly used during design validation, such as code coverage, toggle coverage and fault coverage. Unfortunately, these measures do not have any direct relationship to the functionality of the device. For example, none of these determine if all possible interactions of hazards, stalls and exceptions are tested in a processor pipeline. There is a need for a coverage metric based on the functionality of the design. To define a useful functional coverage metric, we need to define a fault model of the design that is described at the functional level and independent of the implementation details.

In this section, we present a functional fault model for pipelined processors. The fault model should be applicable to a wide variety of today's microprocessors from various architectural domains. We have developed a graph-theoretic model (Section 3.1.1) that can capture a wide spectrum of pipelined processors, coprocessors, and heterogeneous memory subsystems. We have defined functional coverage based on the effects of faults in the fault model applied at the level of the graph-theoretic model. This allows us to compute functional coverage of a pipelined processor for a given set of random or constrained-random test sequences. We present test generation procedures that accept the graph model of the pipelined processor as input and generate test programs to detect all the faults in the functional fault model.

6.2.1 Functional Fault Models

The universe of design errors consists of many types of faults including functional (logical) faults that affect the logic function, and timing faults that effect the operating speed of the system. We only consider the functional faults. The set of possible functional faults (bugs) is dependent on the functionality of the design. In this section, we present fault models for various functions in a pipelined processor. We categorize various computations in a pipelined processor into *register read/write*, *operation execution*, *execution path* and *pipeline execution*. We outline the underlying fault mechanisms for each fault model, and describe the effects of these faults at the level of the graph-based architecture model presented in Section 3.1.1.

Fault Model for Register Read/Write

To ensure fault-free execution, all registers should be written and read correctly. In the presence of a fault, reading of a register will not return the previously written value. The fault could be due to an error in reading, register decoding, register storage, or prior writing. The outcome is an unexpected value. If V_{R_i} is written in register R_i and read back, the output should be V_{R_i} in fault-free case. In the presence of a fault, the output is not equal to V_{R_i}.

Fault Model for Operation Execution

All operations must execute correctly if there are no faults. In the presence of a fault, the output of the computation is different from the expected output. The fault could be due to an error in operation decoding, control generation or final computation. Erroneous operation decoding might return an incorrect opcode. This can happen if incorrect bits are decoded for the opcode. Selection of incorrect bits will also lead to erroneous decoding of source and destination operands. Even if the decoding is correct, due to an error in control generation an incorrect computation unit can be enabled. Finally, the computation unit can be faulty. The outcome is an unexpected result.

Let val_i, where $val_i = f_{opcode_i}(src_1, src_2, ...)$, denote the result of computing the operation "$opcode_i$ $dest$, src_1, src_2, ...". In the fault-free case, the destination will contain the value val_i. Under a fault, the content of the destination is not equal to val_i.

Fault Model for Execution Path

During execution of an operation in the pipeline, one pipeline path and one or more data-transfer paths are activated[1]. We define all these activated paths as the *execution path* for that operation. An execution path ep_{op_i} is faulty if it produces an incorrect result during execution of operation op_i in the pipeline. The fault could be due to an error in one of the paths (pipeline or data-transfer) in the execution path. A path is faulty if any one of its nodes or edges are faulty. A node is faulty if it accepts valid inputs and produces incorrect outputs. An edge is faulty if it does not transfer the data/instruction correctly.

Without loss of generality, let us assume that the processor has p pipeline paths ($PP = \cup_{i=1}^{p} pp_i$) and q data-transfer paths ($DP = \cup_{j=1}^{q} dp_j$). Furthermore, each pipeline path pp_i is connected to a set of data-transfer paths $DPgrp_i$ ($DPgrp_i \subseteq DP$). During execution of an operation op_i in the pipeline path pp_i, a set of data-transfer paths DP_{op_i} ($DP_{op_i} \subseteq DPgrp_i$) are used (activated). Therefore, the execution path ep_{op_i} for operation op_i is, $ep_{op_i} = pp_i \cup DP_{op_i}$. Let us assume, operation op_i has one opcode ($opcode_i$), m sources ($\cup_{j=1}^{m} src_j$) and n destinations ($\cup_{k=1}^{n} dest_k$). Each data-transfer path dp_i ($dp_i \in DP_{op_i}$) is activated to read one of the sources or write one of the destinations of op_i in execution path ep_{op_i}.

[1]Pipeline and data-transfer paths are described in Section 3.1.1

Let val_i, where $val_i = f_{opcode_i}(\cup_{j=1}^{m} src_j)$, denote the result of computing the operation op_i in execution path ep_i. The val_i has n components $(\cup_{k=1}^{n} val_i^k)$. In the fault-free case, the destinations will contain correct values, i.e., $\forall k\ dest_k = val_i^k$. Under a fault, at least one of the destinations will have incorrect value, i.e., $\exists k\ dest_k \neq val_i^k$.

Fault Model for Pipeline Execution

The previous fault models consider only one operation at a time. An implementation of a pipeline is faulty if it produces incorrect results due to execution of multiple operations in the pipeline. The fault could be due to incorrect implementation of the pipeline controller. The faulty controller might have erroneous hazard detection, incorrect stalling, erroneous flushing, or wrong exception handling schemes.

Let us define stall set for a unit u (SS_u) as all possible ways to stall that unit. Therefore, the stall set for the architecture $StallSet = \cup_{\forall u} SS_u$. Let us also define exception set for a unit u (ES_u) as all possible ways to create an exception in that unit. We define the set of all possible multiple exception scenarios as $MESS$. Hence, the exception set for the architecture $ExceptionSet = \cup_{\forall u} ES_u \cup MESS$. We consider two types of pipeline interactions: stalls and exceptions. Therefore, all possible pipeline interactions (PIs) can be defined as: $PIs = StallSet \cup ExceptionSet$. Let us assume a sequence of operations ops_{pi} causes a pipeline interaction pi (i.e., $pi \in PIs$), and updates n storage locations.

Let val_{pi} denote the result of computing the operation sequence ops_{pi}. The val_{pi} has n components $(\cup_{k=1}^{n} val_{pi}^k)$. In the fault-free case, the destinations will contain correct values, i.e., $\forall k\ dest_k = val_i^k$. Under a fault, at least one of the destinations will have incorrect value, i.e., $\exists k\ dest_k \neq val_i^k$

6.2.2 Functional Coverage Estimation

We define functional coverage based on the fault models described in Section 6.2.1.

- a fault in *register read/write* is covered if the register is written first and read later.

- a fault in *operation execution* is covered if the operation is performed, and the result of the computation is read.

- a fault in *execution path* is covered if the execution path is activated, and the result of the computation is read.

- a fault in *pipeline execution* is covered if the fault is activated due to execution of multiple operations in the pipeline, and the result of the computation is read.

We compute functional coverage of a pipelined processor using the traditional definition of coverage. The functional coverage for a given set of test programs is computed as the ratio between the number of faults detected by the test programs and the total number of detectable faults in the fault model.

6.2.3 Test Generation Techniques

In this section, we present test generation procedures for detecting faults covered by the fault models presented in Section 6.2.1. Different architectures have specific instructions to observe the contents of the registers and memories. In our framework, we use load and store instructions to make the register and memory contents observable at the output data bus.

Procedure 1: *createTestProgram*
Input: An operation list *operList*.
Output: Modified operation list with initializations.
begin
    resOperations = {};
    for each operation *oper* in *operList*
        if there are unspecified fields in *oper*
            assign appropriate opcode/operands;
        endif
        for each source *src* of *oper*
            if (*src* is a register or memory location) then
                initOper: initialize *src* with appropriate value;
                resOperations = *resOperations* ∪ *initOper*;
            endif
        endfor
        resOperations = *resOperations* ∪ *oper*;
        readOper: create an operation to read the destination of *oper*;
        resOperations = *resOperations* ∪ *readOper*;
    endfor
    return *resOperations*.
end

We first describe a procedure *createTestProgram* (Procedure 1) that is used by the test generation algorithms. Procedure 1 accepts a list of operations as input and returns the modified list of operations. First, it assigns appropriate values to the unspecified locations (opcodes or operands). Next, it creates initialization instructions for the uninitialized source operands. It also creates instructions to read the destination operands. Finally, it returns the modified list that contains the initialization operations, modified input operations, and the read operations (in that order).

Consider an input list with one operation *ADD dest/reg R1 src2/imm*. The operation has two unspecified fields: *dest* and *src2*. Procedure 1 assigns a register *R3* to *dest* field and an immediate value to *src2* field. It also creates an initialization operation for the source *R1*. Finally, it creates an operation to read the destination. The modified list consists of three operations (in that order): *MOVI R1* 0×5, *ADD R3 R1* 0×23, and *STORE R3 R6* 0×0.

Test Generation for Register Read/Write

Algorithm 8 presents the procedure for generating test programs for detecting faults in register read/write functions. The fault model for the register read/write function is described in Section 6.2.1. For each register in the architecture, the algorithm generates an instruction sequence consisting of a write followed by a read for that register. The function *GenerateUniqueValue* returns unique value for each register based on the register name. A test program for register R_i will consist of two assembly instructions: "MOVI R_i, $\#val_i$" and "STORE R_i, R_j, 0×0". The move-immediate (MOVI) instruction writes val_i in register R_i. The STORE instruction reads the content of R_i and writes it in memory addressed by R_j (offset 0).

Algorithm 8: *Test Generation for Register Read/Write*
Input: Graph model of the architecture *G*.
Output: Test programs for detecting faults in register read/write.
begin /*** TestProgramList = {} ***/
  for each register *reg* in architecture *G*
    $value_{reg}$ = GenerateUniqueValue(*reg*);
    writeInst = an instruction that writes $value_{reg}$ in register *reg*.
    $testprog_{reg}$ = createTestProgram(*writeInst*)
    TestProgramList = *TestProgramList* $\cup$ $testprog_{reg}$;
  endfor
  return *TestProgramList*.
end

Theorem 6.2.1. *The test sequence generated using Algorithm 8 is capable of detecting any detectable fault in the register read/write fault model.*

Proof. Algorithm 8 generates one test program for each register in the architecture. A test program consists of two instructions - a write followed by a read. Each register is written with a specific value. If there is a fault in register read/write function, the value read would be different than the written value. □

Test Generation for Operation Execution

Algorithm 9 presents the procedure for generating test programs for detecting faults in operation execution. The fault model for the operation execution is described in Section 6.2.1. The algorithm traverses the behavior graph of the architecture, and generates one test program for each operation graph using *createTestProgram*. For example, a test program for the operation graph with opcode *ADD* in Figure 3.3 has four operations: two initialization operations ("MOV R3 0×333", "MOV R5 0×212") followed by the ADD operation ("ADD R2 R3 R5"), followed by the reading of the result ("STORE R2, R7, 0×0").

Algorithm 9: *Test Generation for Operation Execution*
Input: Graph model of the architecture G.
Output: Test programs for detecting faults in operation execution.
begin /*** TestProgramList = {} ***/
  for each operation *oper* in architecture G
    $testprog_{oper}$ = createTestProgram(*oper*);
    $TestProgramList = TestProgramList \cup testprog_{oper}$;
  endfor
  return *TestProgramList*.
end

Theorem 6.2.2. *The test sequence generated using Algorithm 9 is capable of detecting any detectable fault in the operation execution fault model.*

Proof. Algorithm 9 generates one test program for each operation in the architecture. If there is a fault in operation execution, the computed result would be different than the expected output. □

Test Generation for Execution Path

Algorithm 10 presents the procedure for generating test programs for detecting faults in execution path. The fault model for the execution path is described in Section 6.2.1. The algorithm traverses the structure graph of the architecture, and for each pipeline path it generates a group of operations supported by that path. It randomly selects one operation from each operation group. There are two possibilities. If all the edges in the execution path (containing the pipeline path) are activated by the selected operation, the algorithm generates all possible source/destination assignments for that operation. However, if different operations in the operation group activate different set of edges in the execution path, it generates all possible source/destination assignments for each operation in the operation group.

Algorithm 10: *Test Generation for Execution Path*
Input: Graph model of the architecture *G*.
Output: Test programs for detecting faults in execution path.
begin /*** TestProgramList = {} ***/
    for each pipeline path *path* in architecture *G*
        $opgroup_{path}$ = operations supported in *path*.
        $exec_{path}$ = *path* and all data-transfer paths connected to it
        $oper_{path}$ = randomly select an operation from $opgroup_{path}$
        if ($oper_{path}$ activates all edges in $exec_{path}$)
            $ops_{path} = oper_{path}$
        else
            $ops_{path} = opgroup_{path}$
        endif
        for all operations *oper* in ops_{path}
            for all source/destination operands *opnd* of *oper*
                for all possible register values *val* of *opnd*
                    newOper = assign *val* to *opnd* of *oper*.
                    $testprog_{oper}$ = createTestProgram(newOper).
                    TestProgramList = *TestProgramList* ∪ $testprog_{oper}$;
                endfor
            endfor
        endfor
    endfor
    return *TestProgramList*.
end

Theorem 6.2.3. *The test sequence generated using Algorithm 10 is capable of detecting any detectable fault in the execution path fault model.*

Proof. The proof is by contradiction. The only way a detectable fault will be missed if a pipeline or data-transfer edge is not activated (used) by the generated test programs. Let us assume, an edge e_{pp} is not activated by any operation. If the e_{pp} is not part of (connected to) any pipeline path, the fault is not detectable. Let us further assume, e_{pp} is part of pipeline path pp. If the pipeline path e_{pp} does not support any operations, the fault is not detectable. If it does support operations, Algorithm 10 will generate operation sequences that exercises this pipeline path and all the data-transfer paths connected to it. Since, the edge e_{pp} is connected to pipeline path pp, it is activated. □

Test Generation for Pipeline Execution

Algorithm 11 presents the procedure for generating test programs for detecting faults in pipeline execution. The fault model for the pipeline execution is described in Section 6.2.1. The first loop (L1) traverses the structure graph of the architecture in a bottom-up manner, starting at leaf nodes. The second loop (L2) computes test programs for generating all possible exceptions in each unit using templates. The third loop (L3) computes test programs for creating stall conditions due to data and control hazards in each unit using templates. The fourth loop (L4) creates test programs to generate stall conditions using structural hazards. Finally, the last loop (L5) computes test sequences for multiple exceptions involving more than one units. The *composeTestProgram* function uses ordered[2] n-tuple units and combines their test programs. The function also removes dependencies across test programs to ensure generation of multiple exceptions during execution of the combined test program.

Theorem 6.2.4. *The test sequence generated using Algorithm 11 is capable of detecting any detectable fault in the pipeline execution.*

Proof. Algorithm 11 generates test programs for all possible interactions during pipeline execution. The first *for* loop (L1) generates all possible hazard and exception scenarions for each functional unit in the pipeline. The test programs for creating all possible exceptions in each node are generated by the second *for* loop (L2). The third *for* loop (L3) generates test programs for creating all possible data and control hazards in each node. Similarly, the fourth *for* loop (L4) generates tests

[2]The unit closer to completion has higher order.

for creating all possible structural hazards in a node. Finally, the last *for* loop (L5) generates test programs for creating all possible multiple exception scenarios in the pipeline. □

Algorithm 11: *Test Generation for Pipeline Execution*
Input: Graph model of the architecture G.
Output: Test programs for detecting faults in pipeline execution.
begin /*** TestProgramList = {} ***/
    L1: **for** each unit node *unit* in architecture G
        L2: **for** each exception *exon* possible in *unit*
            $template_{exon}$ = template for exception *exon*
            $testprog_{unit}$ = createTestProgram($template_{exon}$);
            $TestProgramList = TestProgramList \cup testprog_{unit}$;
        endfor
        L3: **for** each hazard *haz* in {RAW, WAW, WAR, control}
            $template_{haz}$ = template for hazard *haz*
            if *haz* is possible in *unit*
                $testprog_{unit}$ = createTestProgram($template_{haz}$);
                $TestProgramList = TestProgramList \cup testprog_{unit}$;
            endif
        endfor
        L4: **for** each parent unit *parent* of *unit*
            $oper_{parent}$ = an operation supported by *parent*
            resultOps = createTestProgram($oper_{parent}$);
            $testprog_{unit}$ = a test program to stall *unit* (if exists)
            $testprog_{parent} = resultOps \cup testprog_{unit}$
            $TestProgramList = TestProgramList \cup testprog_{parent}$;
        endfor
    endfor
    L5: **for** each ordered n-tuple ($unit_1$, $unit_2$, ..., $unit_n$) in graph G
        $prog_1$ = a test program for creating exception in $unit_1$
        .....
        $prog_n$ = a test program for creating exception in $unit_n$
        $testprog_{tuple}$ = composeTestProgram($prog_1 \cup ... \cup prog_n$);
        $TestProgramList = TestProgramList \cup testprog_{tuple}$;
    endfor
    return *TestProgramList*.
end

6.2.4 A Case Study

We have applied our methodology on two pipelined architectures: a VLIW implementation of the DLX architecture [55], and a RISC implementation of the SPARC V8 architecture [52].

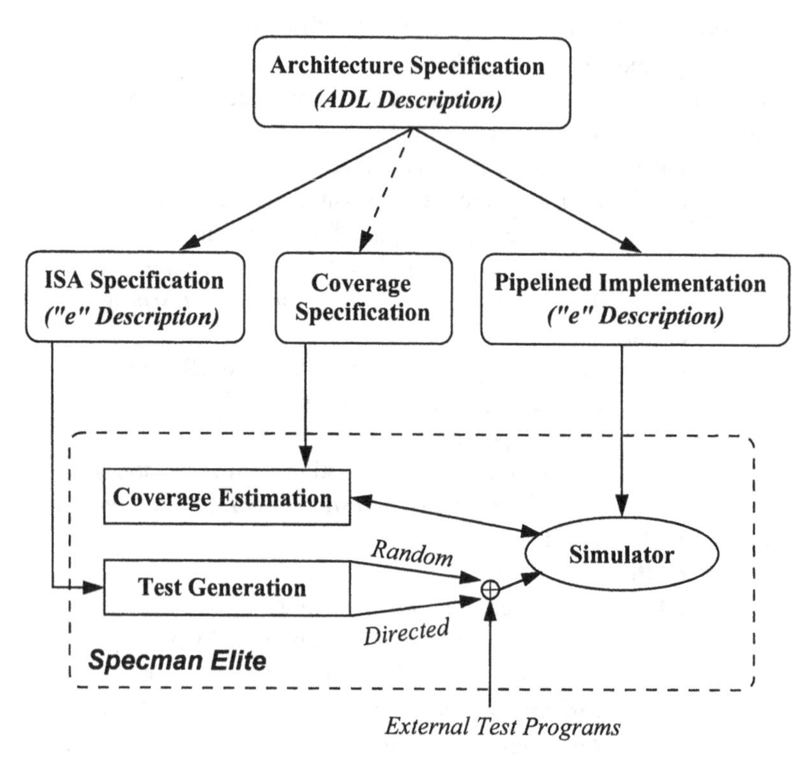

Figure 6.3: Test Generation and Coverage Estimation

Experimental Setup

We have developed our test generation and coverage analysis framework using Verisity's Specman Elite [134] as shown in Figure 6.3. We have captured executable specification of the architectures using Verisity's "e" language. This includes description of 91 instructions for the DLX, and 106 instructions for the SPARC V8 architecture. We refer these as *specifications*. We have implemented a VLIW version of the DLX architecture (shown in Figure 3.6) using Verisity's "e" language. We have used the LEON2 processor [70] that is a VHDL model of a 32-

bit processor compliant with the SPARC V8 architecture. We refer these models (VLIW DLX and LEON2) as *implementations*.

Our framework generates test programs in three different ways: random, constrained random, and our approach. Specman Elite [134] is used to generate both random and constrained-random test programs from the specification. Several constraints are used for constrained-random test generation. For example, we have used the highest probability for choosing register-type operations in DLX to generate test programs for register read/write. Since, register-type operations have three register operands, the chances of reading/writing registers are higher than immediate type (two register operands) or branch type (one register operand) operations. The test programs generated by our approach uses the algorithms described in Section 6.2.3.

To ensure that the generated test programs are executed correctly, our framework applies the test programs on the implementation as well as the specification, and compares the contents of the program counter, registers and memory locations after execution of each test program as shown in Figure 6.4.

The Specman Elite framework allows definition of various coverage measures that enables us to compute the functional coverage described in Section 6.2.2. We defined each entry in the instruction definition (e.g. opcode, destination and sources) as a coverage item in Specman Elite. The coverage for the destination operand gives the measure of which registers are written. Similarly, the coverage of source operands gives the measure of which registers are read. We have used a variable for each register to identify a read after a write. Computation of coverage for operation execution is done by observing the coverage of the opcode field. The computation of coverage for execution path is performed by observing if all the registers are used for computation of all/selected opcodes. This is performed by using cross coverage of instruction fields in Specman Elite that computes every combination of values of the fields. Finally, we compute the coverage for pipeline execution by maintaining variables for stalls and exceptions in each unit. The coverage for multiple exceptions is obtained by performing cross coverage of the exception variables (events) that occur simultaneously.

Results

In this section, we compare the test programs generated by our approach against the random and constrained-random test programs generated by the Specman Elite.

Table 6.3 shows the comparative results for the DLX architecture. The rows indicate the fault models, and the columns indicate test generation techniques. An entry in the table has two numbers. The first one represents the number of operations generated by that test generation technique for that fault model. The second

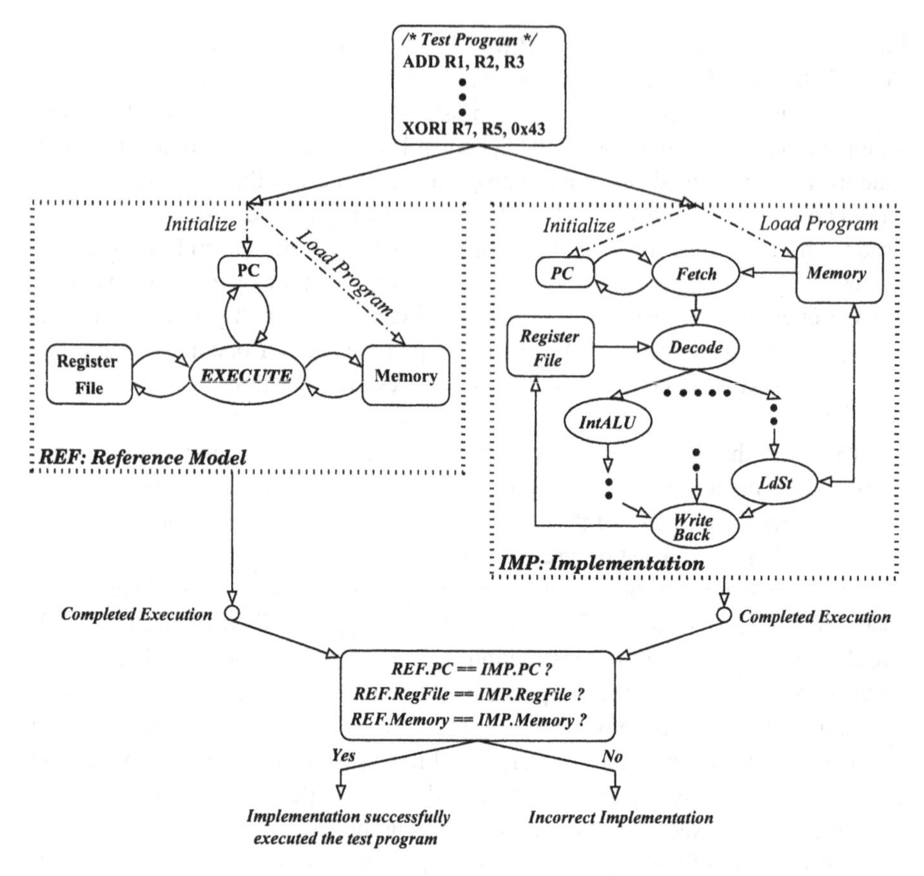

Figure 6.4: Validation of the Implementation

number (in parenthesis) represents the functional coverage obtained by the generated test programs for that fault model. The number 100% implies that the generated test programs covered all the faults in that fault model. For example, the *Random* technique covered all the faults in "*Register Read/Write*" function using 3900 tests. The number of test programs for operation execution are similar for both random and constrained-random approaches. This is because the constraint used in this case (same probability for all opcodes) may be the default option used in random test generation approach.

We performed an initial study to evaluate the quality of our functional fault model using existing coverage measures. Table 6.4 compares our functional coverage against HDL code coverage. The first column indicates the functional fault models. The second column presents the minimum number of test programs gen-

Table 6.3: Test programs for validation of DLX architecture

Fault Models	Test Generation Techniques		
	Random	Constrained	Our Approach
Register Read/Write	3900 (100%)	750 (100%)	130 (100%)
Operation Execution	437 (100%)	443 (100%)	182 (100%)
Execution Path	12627 (100%)	1126 (100%)	320 (100%)
Pipeline Execution	30000 (25%)	30000 (30%)	626 (100%)

erated by our test generation algorithms to cover all the functional faults in the corresponding fault model. The last column presents the code coverage obtained for the DLX implementation [139] using the test programs mentioned in the second column. As expected, our fault model performed well – a small number of test programs generated a high code coverage.

Table 6.4: Quality of the proposed functional fault model

Fault Models	Test Programs	HDL Code Coverage
Register Read/Write	130	85%
Operation Execution	182	91%
Execution Path	320	86%
Pipeline Execution	626	100%

Table 6.5 shows the comparative results for different test generation approaches for the LEON2 processor. The trend is similar in terms of number of operations and functional coverage for both the DLX and LEON2 architectures. The random and constrained-random approaches have obtained 100% functional coverage for the first three fault models using an order of magnitude more test vectors than our approach. We have analyzed the cause for the low functional coverage in *pipeline execution* for the random and constraint-driven test generation approaches. These two approaches covered all the stall scenarios and majority of the single exception faults. However, they could not activate any multiple exception scenarios. Due to the bigger pipeline structure (larger set of pipeline interactions) in the VLIW DLX, it has lower fault coverage than the LEON2 architecture (5-stage integer pipeline) in *pipeline execution*.

Our test generation and coverage estimation framework for the DLX processor is available as an *eShare* (user contributed "e" solutions) in Verisity verification vault [135]. This includes "e" specifications for both ISA description and pipelined implementation of the DLX architecture. It also includes components for random

Table 6.5: Test programs for validation of LEON2 processor

Fault Models	Test Generation Techniques		
	Random	Constrained	Our Approach
Register Read/Write	1746 (100%)	654 (100%)	130 (100%)
Operation Execution	416 (100%)	467 (100%)	212 (100%)
Execution Path	1500 (100%)	475 (100%)	192 (100%)
Pipeline Execution	30000 (40%)	30000 (50%)	248 (100%)

and constrained-random test generation as well as interface for incorporating external tests during simulation. Finally, it includes components for data and temporal checking, and functional coverage estimation.

6.3 Related Work

Traditionally, validation of a microprocessor has been performed by applying a combination of random and directed test programs using simulation techniques. Aharon et al. [1] have proposed a test program generation methodology for functional verification of PowerPC processors in IBM. Miyake et al. [59] have presented a combined scheme of random test generation and specific sequence generation. A coverage driven test generation technique is presented by Fine et al. [119]. Shen et al. [63] have used the processor to generate tests at run-time by self-modifying code, and performed signature comparison with the one obtained from emulation.

Ur and Yadin [123] presented a method for generation of assembler test programs that systematically probe the micro-architecture of a PowerPC processor. Iwashita et al. [37] use an FSM based processor modeling to automatically generate test programs. Campenhout et al. [19] have proposed a test generation algorithm that integrates high-level treatment of the datapath with low-level treatment of the controller. Kohno et al. [66] have presented a tool that generates test programs for verifying pipeline behavior in the presence of hazards and exceptions. Ho et al. [114] have presented a technique for generating test vectors for verifying the corner cases of the design.

Many researchers have proposed techniques for generation of functional test programs for manufacturing testing of microprocessors ([68], [26], [122]). These techniques use stuck-at fault coverage to demonstrate the quality of the generated tests. The applicability of these test programs are not shown for functional validation of microprocessors.

6.4 Chapter Summary

Specification-driven test program generation is a promising approach for functional validation of pipelined processors. In this chapter, we presented two test generation techniques. The first half of the chapter presented a model checking based functional test program generation technique for pipelined processors. Our methodology accepts an ADL specification of the processor as input. A graph model of the pipelined processor is generated from the ADL specification. We defined the functional coverage of the pipeline behavior in terms of the graph coverage. We presented a test program generation algorithm that traverses the pipeline graph to generate test programs based on the coverage metric. Our technique reduced the test generation time and the required BDD size by an order of magnitude.

The second half of the chapter presented a functional coverage based test generation technique for pipelined architectures. The methodology made two important contributions. First, we presented a functional fault model that is used in defining the functional coverage. Second, we presented test generation procedures that accept the graph model of the microprocessor as input and generate test programs to detect all the faults in the functional fault model. We are able to measure the goodness of a given set of random test sequences using our functional coverage metric. We applied this technique on two pipelined architectures: DLX and LEON2. Our experimental results demonstrate that the required number of test sequences generated by our algorithms to obtain a given fault (functional) coverage is an order of magnitude less than the random or constrained-random test programs.

5.4 Chapter Summary

[illegible]

Part IV

Future Directions

7

CONCLUSIONS

Functional validation is one of the most important problems in today's SOC design methodology. A significant bottleneck in the validation of programmable architectures is the lack of a golden reference model. As a result, many existing approaches employ a bottom-up validation approach by using a combination of simulation techniques and formal methods. This book presented a top-down validation methodology for programmable architectures that complements the existing bottom-up techniques. This chapter draws the conclusions from the research results obtained, and looks at some future work on top-down validation and related issues.

7.1 Research Contributions

This book investigated several issues related to top-down validation of programmable architectures consisting of processor core, coprocessors, and memory subsystem. There are four important problems to be addressed in a specification-driven validation methodology:

- **Specification**: How to capture a wide variety of programmable architectures using a specification language? The language should be powerful enough to specify the wide spectrum of contemporary processor, coprocessor, and memory features. On the other hand, the language should be simple enough to allow correlation of the information between the specification and the architecture manual.

- **Specification Validation**: How to validate the architecture specification to ensure it is golden? Specification analysis and validation would be an easier task if the specification language has formal semantics.

- **Model Generation**: How to generate hardware, simulation models, and models for other validation techniques from the given specification?

- **Design Validation**: What are the bottom-up validation techniques that the top-down methodology can complement?

This book examined all of the problems mentioned above. We used the EXPRESSION ADL [5] to specify the architecture. It can capture the structure and behavior of a wide variety of programmable architectures including RISC, DSP, VLIW, and superscalar. The validation techniques we developed are applicable to any specification language that captures both the structure and the behavior of the architecture.

We developed validation techniques to ensure that the static behavior of the pipeline is well-formed by analyzing the structural aspects of the specification using a graph based model. The dynamic behavior is verified by analyzing the instruction flow in the pipeline using a FSM-based model to validate several architectural properties such as determinism and in-order execution in the presence of hazards and multiple exceptions. These properties are by no means complete to prove the correctness of the specification. The designer can add new architecture specific properties and easily integrate it in our validation framework.

A major challenge in top-down validation methodology is the ability to generate executable models from the specification for a wide variety of programmable architectures. We defined a functional abstraction technique to enable generation of models for simulation, hardware generation, and property checking from the ADL specification. The generated simulation and hardware models are used for functional validation and design space exploration of programmable architectures.

This book explored two top-down validation scenarios that complement existing bottom-up techniques: design validation and test generation. The generated hardware is used as a reference model for verifying the hand-written RTL implementation using a combination of symbolic simulation and equivalence checking. We also developed specification-driven test generation techniques based on the functional coverage of the pipelined architectures.

7.2 Future Directions

Top-down validation of programmable architectures will continue to be a major problem. There are many challenges remaining to make this approach viable in practice. The work presented in this book can be extended in the following directions:

- ADLs allow ease of specification for programmable architectures. Formal languages allow specification in a rigorous form. An interesting direction is to develop a specification language that combines the benefits of both.

- There are two important problems that needs to be investigated during specification validation. First, it is necessary to develop architecture specific properties such as the validation of execution style for an out-of-order superscalar processor. Second, it is important to develop a completeness criteria (to establish both necessary and sufficient conditions) for specification validation.

- The functional abstraction based approach we developed in this book allows model generation for uni-processor architectures. There is a need for a methodology to generate models from the specification of programmable architectures containing multiple processor cores.

- Specification-driven hardware generation and validation of design implementation using generated hardware model has one limitation: the generated hardware model (reference) should have a structure similar to the implementation. The requirement is primarily due to the limitation of the equivalence checkers available today. There is a need for new validation techniques that would enable reference model generation and design validation without any knowledge of the implementation details.

- The generated test programs are applied to a cycle-accurate simulator. It would be interesting to perform functional validation of RTL implementation using the generated test programs. We have investigated the applicability of our technique on two simple pipelined processors: DLX and LEON2. Applicability of these techniques can be investigated on today's microprocessors. It is necessary to perform further comparative studies with our functional coverage metric against existing coverage measures, such as code coverage and stuck-at coverage.

- This book considered programmable architectures consisting of a processor core, coprocessor, and memory subsystem. Traditional embedded systems contain many more components including DMAs, input/output devices, specific hardware elements, buses, and so on. It is necessary to extend the current methodology for specification, model generation, and top-down validation of heterogeneous embedded systems.

Part V

Appendices

SURVEY OF CONTEMPORARY ADLS

Section 2.1 presented an overview of ADLs. Figure 2.2 shows the classification of ADLs based on two aspects: *content* and *objective.* The content-oriented classification is based on the nature of the information an ADL can capture, whereas the objective-oriented classification is based on the purpose of an ADL. This appendix presents a survey using content-based classification of ADLs. There are many comprehensive ADL surveys available in the literature including ADLs for retargetable compilation by Qin et al. [137], ADLs for programmable embedded systems by Mishra et al. [101], and ADLs for SOC design by Tomiyama et al. [38].

A.1 Structural ADLs

The structural ADLs capture the structure in terms of architectural components and their connectivity. Early ADLs are based on register-transfer level descriptions: lower abstraction level to enable detailed modeling of digital systems. This section briefly describes two structural ADLs: MIMOLA [117] and UDL/I [34].

MIMOLA

MIMOLA [117] is a structure-centric ADL developed at the University of Dortmund, Germany. It was originally proposed for micro-architecture design. One of the major advantages of MIMOLA is that the same description can be used for synthesis, simulation, test generation, and compilation. A tool chain including the MSSH hardware synthesizer, the MSSQ code generator, the MSST self-test program compiler, the MSSB functional simulator, and the MSSU RT-level simulator were developed based on the MIMOLA language [117]. MIMOLA has also been used by the RECORD [116] compiler.

MIMOLA description contains three parts: the algorithm to be compiled, the target processor model, and additional linkage and transformation rules. The software part (algorithm description) describes application programs in a PASCAL-like syntax. The processor model describes micro-architecture in the form of a component netlist. The linkage information is used by the compiler in order to locate important modules such as program counter and instruction memory. The following code segment specifies the program counter and instruction memory locations [117]:

```
LOCATION_FOR_PROGRAMCOUNTER PCReg;
LOCATION_FOR_INSTRUCTIONS IM[0..1023];
```

The algorithmic part of MIMOLA is an extension of PASCAL. Unlike other high level languages, it allows references to physical registers and memories. It also allows use of hardware components using procedure calls. For example, if the processor description contains a component named MAC, programmers can write the following code segment to use the multiply-accumulate operation performed by MAC:

```
res := MAC(x, y, z);
```

The processor is modeled as a net-list of component modules. MIMOLA permits modeling of arbitrary (programmable or non-programmable) hardware structures. Similar to VHDL, a number of predefined, primitive operators exists. The basic entities of MIMOLA hardware models are modules and connections. Each module is specified by its port interface and its behavior. The following example shows the description of a multi-functional ALU module [117]:

```
MODULE ALU
    (IN inp1, inp2: (31:0);
     OUT outp: (31:0);
     IN ctrl: (1:0);
    )
CONBEGIN
        outp <- CASE ctrl OF
             0: inp1 + inp2 ;
             1: inp1 - inp2 ;
             2: inp1 AND inp2 ;
             3: inp1 ;
             END;
CONEND;
```

The CONBEGIN/CONEND construct includes a set of concurrent assignments. In the example a conditional assignment to output port *outp* is specified, which depends on the two-bit control input *ctrl*. The netlist structure is formed by connecting ports of module instances. For example, the following MIMOLA description connects two modules: *ALU* and accumulator *ACC*.

```
CONNECTIONS ALU.outp -> ACC.inp
            ACC.outp -> ALU.inp
```

The MSSQ code generator extracts instruction-set information from the module netlist. It uses two internal data structures: connection operation graph (COG) and instruction tree (I-tree). It is a very difficult task to extract the COG and I-trees even in the presence of linkage information due to the flexibility of an RT-level structural description. Extra constraints need to be imposed in order for the MSSQ code generator to work properly. The constraints limit the architecture scope of MSSQ to micro-programmable controllers, in which all control signals originate directly from the instruction word. The lack of explicit description of processor pipelines or resource conflicts may result in poor code quality for some classes of VLIW or deeply pipelined processors.

UDL/I

Unified design language, UDL/I is [34] developed as a hardware description language for compiler generation in COACH ASIP design environment at Kyushu University, Japan. UDL/I is used for describing processors at an RT-level on a per-cycle basis. The instruction-set is automatically extracted from the UDL/I description [35], and then it is used for generation of a compiler and a simulator. COACH assumes simple RISC processors and does not explicitly support ILP or processor pipelines. The processor description is synthesizable with the UDL/I synthesis system [39]. The major advantage of the COACH system is that it requires a single description for synthesis, simulation, and compilation. Designer needs to provide hints to locate important machine states such as program counter and register files. Due to difficulty in instruction-set extraction (ISE), ISE is not supported for VLIW and superscalar architectures.

Structural ADLs enable flexible and precise micro-architecture descriptions. The same description can be used for hardware synthesis, test generation, simulation and compilation. However, it is difficult to extract instruction-set without restrictions on description style and target scope. Structural ADLs are more suitable for hardware generation than retargetable compilation.

A.2 Behavioral ADLs

The difficulty of instruction-set extraction can be avoided by abstracting behavioral information from the structural details. Behavioral ADLs explicitly specify the instruction semantics and ignore detailed hardware structures. Typically, there is a one-to-one correspondence between behavioral ADLs and instruction-set reference manual. This section briefly describes three behavioral ADLs: nML [72], ISDL [31] and Valen-C [6].

nML

nML is an instruction-set oriented ADL proposed at Technical University of Berlin. nML has been used by code generators CBC [3] and CHESS [22], and instruction-set simulators Sigh/Sim [28] and CHECKERS. Currently, CHESS/CHECKERS environment is used for automatic and efficient software compilation and instruction-set simulation [49].

nML developers recognized the fact that several instructions share common properties. The final nML description would be compact and simple if the common properties are exploited. Consequently, nML designers used a hierarchical scheme to describe instruction-set. The instructions are the topmost elements in the hierarchy. The intermediate elements of the hierarchy are partial instructions (PI). The relationship between elements can be established using two composition rules: AND-rule and OR-rule. The AND-rule groups several PIs into a larger PI and the OR-rule enumerates a set of alternatives for one PI. Therefore instruction definitions in nML can be in the form of an and-or tree. Each possible derivation of the tree corresponds to an actual instruction.

To achieve the goal of sharing instruction descriptions, the instruction-set is enumerated by an attribute grammar [60]. Each element in the hierarchy has a few attributes. A non-leaf element's attribute values can be computed based on its children's attribute values. Attribute grammar is also adopted by other ADLs such as ISDL [31] and TDL [21].

The following nML description shows an example of instruction specification [72]: The definition of *numeric_instruction* combines three partial instructions (PI) with the AND-rule: *num_action*, SRC, and DST. The first PI, *num_action*, uses OR-rule to describe the valid options for actions: *add* or *sub*. The number of all possible derivations of *numeric_instruction* is the product of the size of *num_action*, *SRC* and *DST*. The common behavior of all these options is defined in the *action* attribute of *numeric_instruction*. Each option for *num_action* should have its own action attribute defined as its specific behavior, which is referred by the *a.action* line. For example, the following code segment has action description for *add* operation. Object code image and assembly syntax can also be specified in the same hierarchical manner.

```
op numeric_instruction(a:num_action, src:SRC, dst:DST)
action {
   temp_src = src;
   temp_dst = dst;
   a.action;
   dst = temp_dst;
}
op num_action = add | sub
op add()
action = {
   temp_dst = temp_dst + temp_src
}
```

nML also captures the structural information used by instruction-set architecture (ISA). For example, storage units should be declared since they are visible to the instruction-set. nML supports three types of storages: RAM, register, and transitory storage. Transitory storage refers to machine states that are retained only for a limited number of cycles e.g., values on buses and latches. Computations have no delay in nML timing model - only storage units have delay. Instruction delay slots are modeled by introducing storage units as pipeline registers. The result of the computation is propagated through the registers in the behavior specification.

nML models constraints between operations by enumerating all valid combinations. The enumeration of valid cases can make nML descriptions lengthy. More complicated constraints, which often appear in DSPs with irregular instruction level parallelism (ILP) constraints or VLIW processors with multiple issue slots, are hard to model with nML. For example, nML cannot model the constraint that operation *I1* cannot directly follow operation *I0*. nML explicitly supports several addressing modes. However, it implicitly assumes an architecture model which restricts its generality. As a result it is hard to model multi-cycle or pipelined units and multi-word instructions explicitly. A good critique of nML is given in [73].

ISDL

Instruction Set Description Language (ISDL) was developed at MIT and used by the Aviv compiler [120] and GENSIM simulator generator [30]. The problem of constraint modeling is avoided by ISDL with explicit specification. ISDL is mainly targeted towards VLIW processors. Similar to nML, ISDL primarily describes the instruction-set of processor architectures. ISDL consists of mainly five sections:

instruction word format, global definitions, storage resources, assembly syntax, and constraints. It also contains an optimization information section that can be used to provide certain architecture specific hints for the compiler to make better machine dependent code optimizations.

The instruction word format section defines fields of the instruction word. The instruction word is separated into multiple fields each containing one or more subfields. The global definition section describes four main types: tokens, non-terminals, split functions and macro definitions. Tokens are the primitive operands of instructions. For each token, assembly format and binary encoding information must be defined. An example token definition of a binary operand is:

```
Token X[0..1] X_R ival {yylval.ival = yytext[1] - '0';}
```

In this example, following the keyword *Token* is the assembly format of the operand. *X_R* is the symbolic name of the token used for reference. The *ival* is used to describe the value returned by the token. Finally, the last field describes the computation of the value. In this example, the assembly syntax allowed for the token *X_R* is *X0* or *X1*, and the values returned are 0 or 1 respectively.

The value (last) field is to be used for behavioral definition and binary encoding assignment by non-terminals or instructions. Non-terminal is a mechanism provided to exploit commonalities among operations. The following code segment describes a non-terminal named *XYSRC*:

```
Non_Terminal ival XYSRC: X_D {$$ = 0;}    |
                         Y_D {$$ = Y_D + 1;};
```

The definition of *XYSRC* consists of the keyword *Non_Terminal*, the type of the returned value, a symbolic name as it appears in the assembly, and an action that describes the possible token or non-terminal combinations and the return value associated with each of them. In this example, *XYSRC* refers to tokens *X_D* and *Y_D* as its two options. The second field (*ival*) describes the returned value type. It returns 0 for *X_D* or incremented value for *Y_D*.

Similar to nML, storage resources are the only structural information modeled by ISDL. The storage section lists all storage resources visible to the programmer. It lists the names and sizes of the memory, register files, and special registers. This information is used by the compiler to determine the available resources and how they should be used.

The assembly syntax section is divided into fields corresponding to the separate operations that can be performed in parallel within a single instruction. For each field, a list of alternative operations can be described. Each operation description consists of a name, a list of tokens or non-terminals as parameters, a set of commands that manipulate the bitfields, RTL description, timing details, and costs. RTL description captures the effect of the operation on the storage resources. Multiple costs are allowed including operation execution time, code size, and costs due to resource conflicts. The timing model of ISDL describes when the various effects of the operation take place.

In contrast to nML, which enumerates all valid combinations, ISDL defines invalid combinations in the form of Boolean expressions. This often leads to a simple constraint specification. It also enables ISDL to capture irregular ILP constraints. The following example shows how to describe the constraint that instruction I1 cannot directly follow instruction I0. The "[1]" indicates a time shift of one instruction fetch for the I0 instruction. The '~' is a symbol for NOT and '&' is for logical AND.

```
~(I1 *) & ([1] I0 *, *)
```

ISDL provides the means for compact and hierarchical instruction-set specification. However, it may not be possible to describe a instruction-set with multiple encoding formats using simple tree-like instruction structure of ISDL.

Valen-C

Valen-C is an embedded software programming language proposed at Kyushu University, Japan [6, 7]. Valen-C is an extended C language which supports explicit and exact bit-width for integer type declarations. A retargetable compiler (called Valen-CC) has been developed that accepts C or Valen-C programs as an input and generates the optimized assembly code. Although Valen-CC assumes simple RISC architectures, it has retargetability to some extent. The most interesting feature of Valen-CC is that the processor can have any datapath bit-width (e.g., 14 bits or 29 bits). The Valen-C system aims at optimization of datapath width. The target processor description for Valen-CC includes the instruction-set consisting of behavior and assembly syntax of each instruction as well as the processor datapath width. Valen-CC does not explicitly support processor pipelines or ILP.

In general, the behavioral languages have one feature in common: hierarchical instruction-set description based on attribute grammar [60]. This feature simplifies the instruction-set description by sharing the common components between operations. However, the capabilities of these models are limited due to the lack of detailed pipeline and timing information. It is not possible to generate cycle accurate simulators without certain assumptions regarding control behavior. Due to lack of structural details, it is also not possible to perform resource-based scheduling using behavioral ADLs.

A.3 Mixed ADLs

Mixed languages captures both structural and behavioral details of the architecture. This section briefly describes four mixed ADLs: FlexWare, HMDES, EXPRESSION, and LISA.

FlexWare

FlexWare is a CAD system for DSP or ASIP design [108]. The FlexWare system includes the CodeSyn code generator and the Insulin simulator. Both behavior and structure are captured in the target processor description. The machine description for CodeSyn consists of three components: instruction-set, available resources (and their classification), and an interconnect graph representing the datapath structure. The instruction-set description is a list of generic processor macro instructions to execute each target processor instruction. The simulator uses a VHDL model of a generic parameterizable machine. The parameters include bit-width, number of registers, ALUs, and so on. The application is translated from the user-defined target instruction-set to the instruction-set of the generic machine. Then, the code is executed on the generic machine.

HMDES

Machine description language HMDES was developed at University of Illinois at Urbana-Champaign for the IMPACT research compiler [54]. C-like preprocessing capabilities such as file inclusion, macro expansion and conditional inclusion are supported in HMDES. An HMDES description is the input to the MDES machine description system of the Trimaran compiler infrastructure, which contains IMPACT as well as the Elcor research compiler from HP Labs. The description is first pre-processed, then optimized and translated to a low-level representation file. A machine database reads the low level files and supplies information for the compiler back end through a predefined query interface.

MDES captures both structure and behavior of target processors. Information is broken down into sections such as format, resource usage, latency, operation, and register. For example, the following code segment describes register and register file. It describes 64 registers. The register file describes the width of each register and other optional fields such as generic register type (virtual field), speculative, static and rotating registers. The value '1' implies speculative and '0' implies non-speculative.

```
SECTION Register {
  R0(); R1(); ... R63();
  'R[0]'(); ... 'R[63]'();
   ...
}

SECTION Register_ File {
  RF_i(width(32) virtual(i) speculative(1)
       static(R0...R63) rotating('R[0]'...'R[63]'));
  ...
}
```

MDES allows only a restricted retargetability of the cycle-accurate simulator to the HPL-PD processor family [80]. MDES permits description of memory systems, but limited to the traditional hierarchy, i.e., register files, caches, and main memory.

EXPRESSION

The above mixed ADLs require explicit description of Reservation Tables (RT). Processors that contain complex pipelines, large amounts of parallelism, and complex storage sub-systems, typically contain a large number of operations and resources (and hence RTs). Manual specification of RTs on a per-operation basis thus becomes cumbersome and error-prone. The manual specification of RTs (for each configuration) becomes impractical during rapid architectural exploration. The EXPRESSION ADL [5] describes a processor as a netlist of units and storages to automatically generate RTs based on the netlist [88]. Unlike MIMOLA, the netlist representation of EXPRESSION is coarse grain. It uses a higher level of abstraction similar to block-diagram level description in architecture manual.

EXPRESSION ADL was developed at University of California, Irvine. The ADL has been used by the retargetable compiler (EXPRESS [4]) and simulator (SIMPRESS [8]) generation framework. The framework also supports a graphical user interface (GUI) and can be used for design space exploration of programmable architectures consisting of processor cores, coprocessors and memories [45].

An EXPRESSION description is composed of two main sections: behavior (instruction-set), and structure. The behavior section has three subsections: operations, instruction, and operation mappings. Similarly, the structure section consists of three subsections: components, pipeline/data–transfer paths, and memory subsystem.

The operation subsection describes the instruction-set of the processor. Each operation of the processor is described in terms of its opcode and operands. The types and possible destinations of each operand are also specified. A useful feature of EXPRESSION is operation group that groups similar operations together for the ease of later reference. For example, the following code segment shows an operation group (*alu_ops*) containing two ALU operations: *add* and *sub*.

```
(OP_GROUP alu_ops
   (OPCODE add
      (OPERANDS (SRC1 reg) (SRC2 reg/imm) (DEST reg))
      (BEHAVIOR DEST = SRC1 + SRC2)
      ...
   )
   (OPCODE sub
      (OPERANDS (SRC1 reg) (SRC2 reg/imm) (DEST reg))
      (BEHAVIOR DEST = SRC1 - SRC2)
      ...
   )
)
```

The instruction subsection captures the parallelism available in the architecture. Each instruction contains a list of slots (to be filled with operations), with each slot corresponding to a functional unit. The operation mapping subsection is used to specify the information needed by instruction selection and architecture-specific optimizations of the compiler. For example, it contains mapping between generic and target instructions.

The component subsection describes each RT-level component in the architecture. The components can be pipeline units, functional units, storage elements, ports, and connections. For multi-cycle or pipelined units, the timing behavior is also specified.

The pipeline/data-transfer path subsection describes the netlist of the processor. The *pipeline path description* provides a mechanism to specify the units which comprise the pipeline stages, while the *data-transfer path description* provides a mechanism for specifying the valid data-transfers. This information is used to both retarget the simulator, and to generate reservation tables needed by the scheduler [88]. An example path declaration for the DLX architecture [55] (Figure 3.6) is shown below. It describes that the processor has five pipeline stages. It also describes that the *Execute* stage has four parallel paths. Finally, it describes each path e.g., it describes that the *FADD* path has four pipeline stages.

```
(PIPELINE Fetch Decode Execute MEM WriteBack)
(Execute (ALTERNATE IALU MULT FADD DIV))
(MULT (PIPELINE MUL1 MUL2 ... MUL7))
(FADD (PIPELINE FADD1 FADD2 FADD3 FADD4))
```

The memory subsection describes the types and attributes of various storage components (such as register files, SRAMs, DRAMs, and caches). The memory netlist information can be used to generate memory aware compilers and simulators [98, 107]. Memory aware compilers can exploit the detailed information to hide the latency of the lengthy memory operations [89].

In general, EXPRESSION captures the data path information in the processor. The control path is not explicitly modeled. Also, the VLIW instruction composition model is simple. The instruction model requires extension to capture inter-operation constraints such as sharing of common fields. Such constraints can be modeled by ISDL through cross-field encoding assignment. Appendix B includes a sample description of the DLX processor shown in Figure 3.6 using EXPRESSION ADL.

LISA

LISA (Language for Instruction Set Architecture) [133] was developed at Aachen University of Technology, Germany with a simulator centric view. The language has been used to produce production quality simulators [121]. An important aspect of LISA language is its ability to capture control path explicitly. Explicit modeling of both datapath and control is necessary for cycle-accurate simulation. LISA has also been used to generate retargetable C compilers [74, 86].

LISA descriptions are composed of two types of declarations: resource and operation. The resource declarations cover hardware resources such as registers, pipelines, and memories. The pipeline model defines all possible pipeline paths that operations can go through. An example pipeline description for the architecture shown in Figure 3.6 is as follows:

```
PIPELINE int = {Fetch; Decode; IALU; MEM; WriteBack}
PIPELINE flt = {Fetch; Decode; FADD1; FADD2;
                FADD3; FADD4; MEM; WriteBack}
PIPELINE mul = {Fetch; Decode; MUL1; MUL2; MUL3; MUL4;
                MUL5; MUL6; MUL7; MEM; WriteBack}
PIPELINE div = {Fetch; Decode; DIV; MEM; WriteBack}
```

Operations are the basic objects in LISA. They represent the designer's view of the behavior, the structure, and the instruction-set of the programmable architecture. Operation definitions capture the description of different properties of the system such as operation behavior, instruction-set information, and timing. These operation attributes are defined in several sections:

- The CODING section describes the binary image of the instruction word.
- The SYNTAX section describes the assembly syntax of instructions.
- The SEMANTICS section specifies the instruction-set semantics.
- The BEHAVIOR section describes components of the behavioral model.
- The ACTIVATION section describes the timing of other operations relative to the current operation.
- The DECLARE section contains local declarations of identifiers.

LISA exploits the commonality of similar operations by grouping them into one. The following code segment describes the decoding behavior of two immediate-type (i_type) operations (ADDI and SUBI) in the DLX *Decode* stage. The complete behavior of an operation can be obtained by combining its behavior definitions in all the pipeline stages.

```
OPERATION i_type IN pipe_int.Decode {
    DECLARE {
        GROUP opcode={ADDI || SUBI}
        GROUP rs1, rd = {fix_register};
    }
    CODING {opcode rs1 rd immediate}
    SYNTAX {opcode rd ``,'' rs1 ``,'' immediate}
    BEHAVIOR { reg_a = rs1; imm = immediate; cond = 0;
    }
    ACTIVATION {opcode, writeback}
}
```

A language similar to LISA is RADL. RADL [15] was developed at Rockwell, Inc. as an extension of the LISA approach that focuses on explicit support of detailed pipeline behavior to enable generation of production quality cycle-accurate and phase-accurate simulators.

A.4 Partial ADLs

The ADLs discussed so far captures complete description of the processor's structure, behavior or both. There are many description languages that captures partial information of the architecture needed to perform a specific task. This section describes two such ADLs.

AIDL

AIDL is an ADL developed at University of Tsukuba for design of high-performance superscalar processors [129]. AIDL aims at validation of the pipeline behavior such as data-forwarding and out-of-order completion. In AIDL, timing behavior of pipeline is described using interval temporal logic. AIDL does not support software toolkit generation. However, AIDL descriptions can be simulated using the AIDL simulator.

PEAS-I

PEAS-I is a CAD system for ASIP design supporting automatic instruction-set optimization, compiler generation, and instruction level simulator generation [61]. In the PEAS-I system, the GNU C compiler is used, and the machine description of GCC is automatically generated. Inputs to PEAS-I include an application program written in C and input data to the program. Then, the instruction-set is automatically selected in such a way that the performance is maximized or the gate count is minimized. Based on the instruction-set, GNU CC and an instruction level simulator is automatically retargeted.

B

SPECIFICATION OF DLX PROCESSOR

This appendix includes a sample description of the DLX processor shown in Figure 3.6 using EXPRESSION ADL.

```
;****************************************************************
;* This machine description is copyrighted by the Regents of the *
;* University of California, Irvine. This is a simplified version*
;* of the EXPRESSION description for the DLX processor. It only  *
;* includes the features that are required for the validation    *
;* steps discussed in this book. No Warranty of any kind.         *
;****************************************************************
;
;**************** Section 1: Specific Operations ****************

(OPERATIONS_SECTION

 (OP_GROUP all
   (OPCODE IADD
     (OP_TYPE DATA_OP)
     (OPERANDS (DEST reg) (SRC1 reg) (SRC2 reg/imm))
     (BEHAVIOR DEST = SRC1 + SRC2)
   )

   (OPCODE ISUB
     (OP_TYPE DATA_OP)
     (OPERANDS (DEST reg) (SRC1 reg) (SRC2 reg/imm))
     (BEHAVIOR DEST = SRC1 - SRC2)
   )

   (OPCODE IMUL
     (OP_TYPE DATA_OP)
     (OPERANDS (DEST reg) (SRC1 reg) (SRC2 reg/imm))
     (BEHAVIOR DEST = SRC1 * SRC2)
   )
```

```
  (OPCODE IDIV
    (OP_TYPE DATA_OP)
    (OPERANDS (DEST reg) (SRC1 reg) (SRC2 reg/imm))
    (BEHAVIOR DEST = SRC1 / SRC2)
  )

  (OPCODE ILSH
    (OP_TYPE DATA_OP)
    (OPERANDS (DEST reg) (SRC1 reg) (SRC2 imm))
    (BEHAVIOR DEST = SRC1 << SRC2)
  )

  (OPCODE ILT
    (OP_TYPE DATA_OP)
    (OPERANDS (DEST reg) (SRC1 reg) (SRC2 reg/imm))
    (BEHAVIOR DEST =  (SRC1 < SRC2) ? 1 : 0)
  )

  (OPCODE IVLOAD
    (OP_TYPE DATA_OP)
    (OPERANDS (DEST reg) (SRC1 reg) (SRC2 imm))
    (BEHAVIOR DEST =  MEMORY[SRC1 + SRC2])
  )

  (OPCODE IVSTORE
    (OP_TYPE DATA_OP)
    (OPERANDS (SRC reg) (SRC1 reg) (SRC2 imm))
    (BEHAVIOR MEMORY[SRC1 + SRC2] = SRC)
  )

  (OPCODE BEQZ
    (OP_TYPE CONTROL_OP)
    (OPERANDS (SRC1 reg) (SRC2 imm))
    (BEHAVIOR PC = (SRC1 == 0) ? SRC2 : PC + 4)
  )

  (OPCODE J
    (OP_TYPE CONTROL_OP)
    (OPERANDS (SRC reg/imm))
    (BEHAVIOR PC = SRC)
  )

  .......
)
```

```
 (OP_GROUP ALU_instr
   (OPCODE ICONSTANT IASSIGN ASSIGN IADD ISUB ILSH IASH IVLOAD
           DVLOAD FVLOAD IVSTORE DVSTORE FVSTORE ILAND IEQ INE
           ILE IGE ILT IGT J BEQZ BNEZ NOP
   )
 )

 (OP_GROUP MUL_instr
  (OPCODE IMUL DMUL FMUL NOP)
 )

 (OP_GROUP FADD_instr
   (OPCODE DCONSTANT FCONSTANT DASSIGN FASSIGN DADD FADD DSUB FSUB
       DEQ FEQ DNE FNE DLE FLE DGE FGE DLT FLT DGT FGT CVTDI CVTSI
       CVTSD CVTDS DMTC1 DMFC1 DNEG MFC1 MTC1 TRUNCID TRUNCIS NOP
   )
 )

 (OP_GROUP FDIV_instr
  (OPCODE IDIV DDIV FDIV NOP)
 )

 (OP_GROUP LDST_instr
  (OPCODE IVLOAD DVLOAD FVLOAD IVSTORE DVSTORE FVSTORE)
 )
)

;***************** Section 2: Instruction template ***************

(INSTRUCTION_SECTION
  (WORDLEN 32)
  (SLOTS
   ((TYPE DATA) (BITWIDTH 8) (UNIT IALU))
   ((TYPE DATA) (BITWIDTH 8) (UNIT MUL))
   ((TYPE DATA) (BITWIDTH 8) (UNIT FADD))
   ((TYPE DATA) (BITWIDTH 8) (UNIT FDIV))
   ((TYPE CONTROL) (BITWIDTH 8) (UNIT DECODE))
  )
)

;***************** Section 3: Operation Mappings ****************

(OPMAPPING_SECTION
  (
     (GENERIC ((IMUL tmpDest SRC1 SRC2) (IADD DEST tmpDest SRC3)))
     (TARGET (MAC DEST SRC1 SRC2 SRC3))
  )
)
```

```
;************* Section 4: Components Specification ***************

(ARCHITECTURE_SECTION
 (SUBTYPE UNIT FetchUnit DecodeUnit IAluUnit FMulUnit FAddUnit
               FDivUnit MemoryUnit WriteBackUnit)
 (SUBTYPE PORT UnitPort Port)
 (SUBTYPE CONNECTION MemoryConnection  RegConnection)
 (SUBTYPE STORAGE VirtualRegFile VirtualMemory)
 (SUBTYPE LATCH PipelineLatch MemoryLatch)

 (FetchUnit FETCH
  (CAPACITY 1) (TIMING (all 1)) (OPCODES all)
  (LATCHES (OUT FetDecLatch) (IN PCLatch))
 )

 (DecodeUnit DECODE
  (CAPACITY 1) (TIMING (all 1)) (OPCODES all)
  (LATCHES (IN FetDecLatch) (OUT DecALULatch DecM1Latch
                                 DecA1Latch DecFDivLatch))
  (PORTS DecRdPort1 DecRdPort2)
 )

 (IntAluUnit IALU
  (CAPACITY 1) (TIMING (all 1)) (OPCODES ALU_instr)
  (LATCHES (IN DecIALULatch) (OUT ExMemLatch))
 )

 (FMultUnit M1
  (CAPACITY 1) (TIMING (all 1)) (OPCODES MUL_instr)
  (LATCHES (IN DecM1Latch) (OUT M1M2Latch))
 )

 (FAddUnit A1
  (CAPACITY 1) (TIMING (all 1)) (OPCODES FADD_instr)
  (LATCHES (IN DecA1Latch) (OUT A1A2Latch))
 )

 (FDivUnit FDIV
  (CAPACITY 1) (OPCODES FDIV_instr)
  (TIMING (IDIV 25) (FDIV 25) (DDIV 25) (NOP 1))
  (LATCHES (IN DecFDivLatch) (OUT ExMemLatch))
 )

 (FMultUnit M2
  (CAPACITY 1) (TIMING (all 1)) (OPCODES MUL_instr)
  (LATCHES (IN M1M2Latch) (OUT M2M3Latch))
 )
```

```
(FMultUnit M3
 (CAPACITY 1) (TIMING (all 1)) (OPCODES MUL_instr)
 (LATCHES (IN M2M3Latch) (OUT M3M4Latch))
)

(FMultUnit M4
 (CAPACITY 1) (TIMING (all 1)) (OPCODES MUL_instr)
 (LATCHES (IN M3M4Latch) (OUT M4M5Latch))
)

(FMultUnit M5
 (CAPACITY 1) (TIMING (all 1)) (OPCODES MUL_instr)
 (LATCHES (IN M4M5Latch) (OUT M5M6Latch))
)

(FMultUnit M6
 (CAPACITY 1) (TIMING (all 1)) (OPCODES MUL_instr)
 (LATCHES (IN M5M6Latch) (OUT M6M7Latch))
)

(FMultUnit M7
 (CAPACITY 1) (TIMING (all 1)) (OPCODES MUL_instr)
 (LATCHES (IN M6M7Latch) (OUT ExMemLatch))
)

(FAddUnit A2
 (CAPACITY 1) (TIMING (all 1)) (OPCODES FADD_instr)
 (LATCHES (IN A1A2Latch) (OUT A2A3Latch))
)

(FAddUnit A3
 (CAPACITY 1) (TIMING (all 1)) (OPCODES FADD_instr)
 (LATCHES (IN A2A3Latch) (OUT A3A4Latch))
)

(FAddUnit A4
 (CAPACITY 1) (TIMING (all 1)) (OPCODES FADD_instr)
 (LATCHES (IN A3A4Latch) (OUT ExMemLatch))
)

(MemoryUnit MEM
 (CAPACITY 1) (TIMING (all 1)) (OPCODES LDST_instr)
 (LATCHES (IN ExMemLatch) (OUT MemWbLatch))
 (PORTS MemUnitPort)
)
```

```
 (WriteBackUnit WRITEBACK
  (CAPACITY 1) (TIMING (all 1)) (OPCODES all)
  (LATCHES (IN MemWbLatch)) (PORTS WbRegWrPort)
 )

 (PipelineLatch M1M2Latch M2M3Latch M3M4Latch M4M5Latch M5M6Latch
    M6M7Latch A1A2Latch A2A3Latch A3A4Latch ExMemLatch MemWbLatch
    FetDecLatch DecALULatch DecM1Latch DecA1Latch DecFDivLatch
 )
 (MemoryLatch InstLatch)
 (UnitPort DecRdPort1 DecRdPort2 MemUnitPort)
 (RegConnection DecRegCxn1 DecRegCxn2 WbRegCxn, MemoryCxn
     FetMemoryCxn
 )
 (Port RegRdPort1 RegRdPort2 RegWrPort MemRdPort MemRdWrPort)
 (VirtualRegFile REGFILE)
 (VirtualMemory MEMORY)
)

;**** Section 5: Pipeline and Data-transfer paths ****

(PIPELINE_SECTION
 (PIPELINE FETCH DECODE EXECUTE MEM WRITEBACK)
 (EXECUTE (PARALLEL IALU FMUL FADD FDIV))
 (FMUL (PIPELINE M1 M2 M3 M4 M5 M6 M7))
 (FADD (PIPELINE A1 A2 A3 A4))

 (DTPATHS
   (REGFILE DECODE RegRdPort1 DecRegCxn1 DecRdPort1)
   (REGFILE DECODE RegRdPort2 DecRegCxn2 DecRdPort2)
   (WRITEBACK REGFILE WbRegWrPort WbRegCxn RegWrPort)
   (MEM MEMORY MemUnitPort MemoryCxn MemoryRdWrPort)
   (FETCH MEMORY FetchRdPort FetMemoryCxn MemoryRdPort)
 )
)

;**************** Section 6: Memory Hierarchy *****************

(STORAGE_SECTION
  (REGFILE
    (TYPE REGFILE) (WIDTH 32) (SIZE 32)
  )

  (MEMORY
    (TYPE RAM) (WIDTH 8) (SIZE 1024) (ACCESS_TIMES 1)
    (ADDRESS_RANGE (0 1023))
  )
)
```

C

INTERRUPTS & EXCEPTIONS IN ADL

Section 2.3 described how to specify programmable architectures including a processor core, coprocessors, and memory subsystem using EXPRESSION ADL. This appendix describes how to capture exceptions and interrupts using the ADL. It is necessary to capture exceptions and interrupts explicitly in the ADL for various reasons. First, the simulator and hardware generators require this information to accurately generate and handle exceptions. Second, the specification validation techniques use this information to analyze pipeline interactions in the presence of multiple exceptions. For example, we have used this information in Section 3.2 to verify in-order execution of pipelined processor specifications. We classify exceptions into three categories: opcode related exceptions, exceptions related to functional units, and external exceptions. The motivation behind this classification is to enable ease of specification.

Opcode related exceptions

It is appropriate to describe opcode related exceptions and their actions inside the opcode specification. For example, the modified *div* operation contains the exception information as shown in Figure C.1.

Exceptions Related to Functional Units

Functional unit related exceptions are defined in ADL's component specification section. For example, the *Decode* unit shown in Figure 3.2 can issue up to three instructions per cycle. The first one is for the *ALU* pipeline, the second one is for the *FADD* pipeline and the last one is for the coprocessor pipeline. It is an

```
# Behavior: description of instruction-set
......
( opcode div
  (operands (s1 reg) (s2 reg) (dst reg)) (behavior dst = s1 / s2) ...
  (exceptions (if (s2 == 0) throw div_by_zero) ...)
)
```

Figure C.1: Specification of division_by_zero exception

exception if the last instruction is not a coprocessor instruction. The specification of such an exception is described in the *Decode* unit as shown in Figure C.2.

```
# Components specification
......
( DecodeUnit Decode
  (capacity 2) (timing (all 1)) (opcodes all) ...
  (exceptions (if (slot3 opcode != coprocessor_type) throw illegal_slot_instruction) ...)
)
```

Figure C.2: Specification of illegal_slot_instruction exception

External Exceptions

External interrupts can be specified at the processor level. We model a control unit that performs the task of a controller. The control unit is also used to perform stalling and flushing of the processor pipelines as described in Section 4.2.4. We describe external interrupts in the control unit. For example, a machine reset exception can be described in the control unit as shown in Figure C.3. We assume that the *reset* is an external interrupt that is used to generate the internal exception *machine_reset*.

Specification of Interrupts

The mapping between exceptions and interrupts is a *many-to-one* mapping function. A class of exceptions may give rise to one interrupt. In such cases the architecture implementation should ensure that only one exception from that class occurs at a time. In general, one interrupt corresponds to more than one exception. We specify the interrupts and exceptions in the control unit specification. For

```
# Components specification
......
( ControlUnit control
  ......
  (exceptions (if reset throw machine_reset) ...)
)
```

Figure C.3: Specification of machine_reset exception

example, the interrupt *int1* is described in Figure C.4. The interrupt *int1* gets generated due to failures during memory operation, for example, ITLB miss or DTLB miss. It can mask several lower priority interrupts such as *int2* and *int7*.

```
# Components specification
......
( ControlUnit control
  ......
  (interrupt int1
   (exceptions ITLB_miss DTLB_miss ...)
   (masks int2 int7 ...) (behavior ...) ...
  )
)
```

Figure C.4: Specification of interrupts

We model the interrupt handler using a priority table that can accept *n* exception requests and generate only one interrupt per cycle. The multiple exceptions are handled in a simple and uniform manner using interrupt service register (ISR). The length of the ISR is equal to the number of interrupts possible in that architecture. One entry in the ISR corresponds to an interrupt. Control unit defines the class of exceptions that generates a particular interrupt. Each exception sets one particular bit in the ISR of the interrupt handler. Interrupt handler decides the highest priority interrupt using the interrupt priority table. Depending on the masking information the highest priority interrupt masks the appropriate bits in ISR. The process of selecting highest priority interrupt continues until there are no bits set in ISR. The details on specification of exceptions and interrupts are available in [103].

D

VALIDATION OF DLX SPECIFICATION

Chapter 3 presented a framework for validation of both static and dynamic properties in architecture specification. This appendix presents a case study for validation of dynamic properties including determinism and in-order execution in DLX processor specification. Figure D.1 shows the DLX processor pipeline that is obtained from Figure 3.6 by adding pipeline registers (latches).

The structure and behavior of the processor is captured using the EXPRESSION ADL [5]. Based on the discussion in Section 3.2.1, we captured the conditions for stalling, normal flow, branch taken and bubble insertion in the ADL. For example, we captured *CacheMiss* as the external signal for PC unit. For all other units we assumed *all* contribution from children units for stall condition. While capturing normal flow condition for each unit we selected *any* for parent units and *any* for children units. Similarly, for each unit we specified *all* as contribution from parent units and *any* as contribution for children units for bubble insertion. The condition specification for the decode unit (no self contribution) is shown below.

```
(DecodeUnit DECODE
     .........
     (CONDITIONS
          (NF ANY ANY)
          (ST ALL)
          (BI ALL ANY)
          (SELF "")
     )
)
```

Using the ADL description, we automatically generated the equations for flow conditions for all the units [95]. For example, the equation for the stall condition for the decode latch is shown below (using Equation (3.8), and the description of the decode unit shown above):

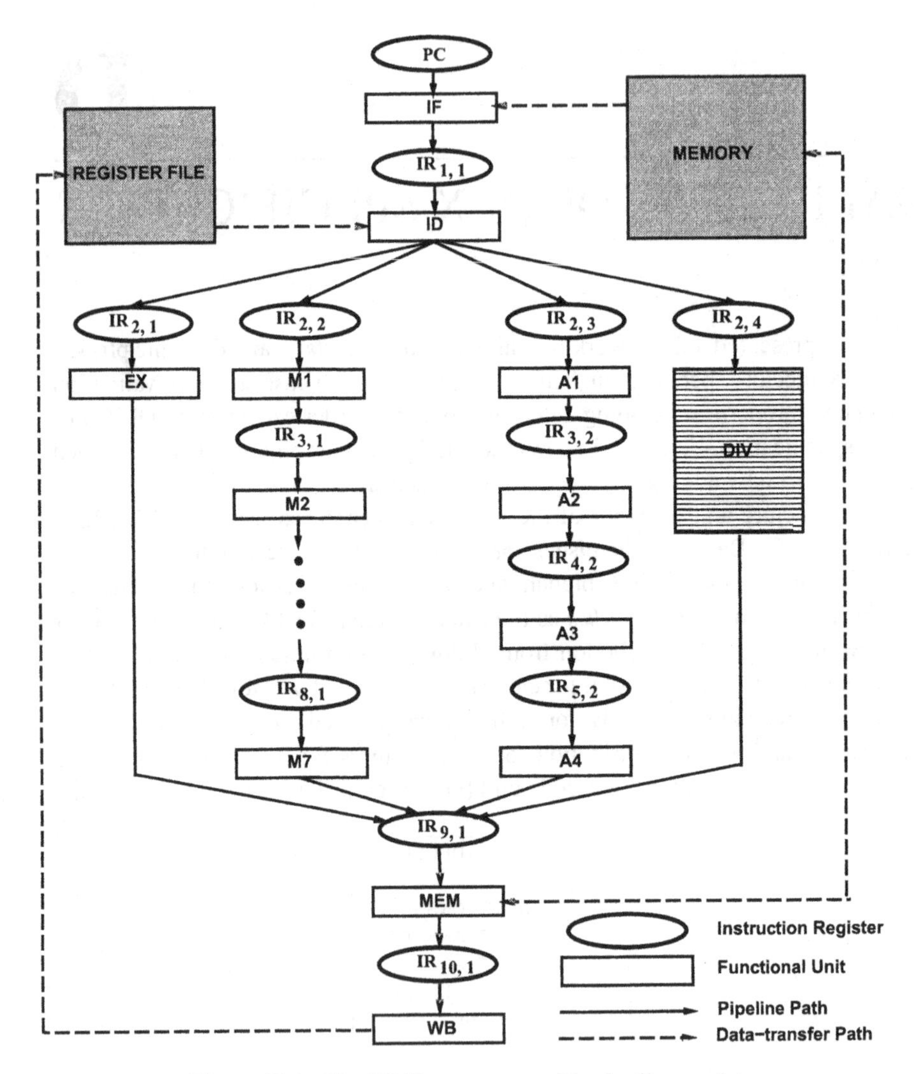

Figure D.1: The DLX processor with pipeline registers

$$cond_{IR_{1,1}}^{ST} = (ST_{IR_{2,1}} \,.\, ST_{IR_{2,2}} \,.\, ST_{IR_{2,3}} \,.\, ST_{IR_{2,4}}).\overline{XN_{IR_{1,1}}}.\overline{SQ_{IR_{1,1}}} \tag{D.1}$$

$IR_{2,4}$ represents latch for the multicycle unit. So we assumed a signal *busy* internal to $IR_{2,4}$ which remained set for *n* cycles. The *busy* can be treated as $ST_{IR_{2,4}}^{self}$ as shown in Equation (3.8).

The necessary equations for verifying the properties such as determinism and in-order execution are generated from the given ADL description. We show here a small trace of the property checking to demonstrate the simplicity and elegance of the underlying model. We show that the determinism property is satisfied for $IR_{1,1}$ using the modeling above:

$$
\begin{aligned}
& cond_{IR_{1,1}}^{NF} + cond_{IR_{1,1}}^{ST} + cond_{IR_{1,1}}^{BI} + cond_{IR_{1,1}}^{XN} + cond_{IR_{1,1}}^{SQ} \\
& = \overline{ST_{PC}} \,.\, (\overline{ST_{IR_{2,1}}} + \overline{ST_{IR_{2,2}}} + \overline{ST_{IR_{2,3}}} + \overline{ST_{IR_{2,4}}}) \,.\, \overline{XN_{IR_{1,1}}} \,.\, \overline{SQ_{IR_{1,1}}} + \\
& (ST_{IR_{2,1}} \,.\, ST_{IR_{2,2}} \,.\, ST_{IR_{2,3}} \,.\, ST_{IR_{2,4}}) \,.\, \overline{XN_{IR_{1,1}}} \,.\, \overline{SQ_{IR_{1,1}}} + \\
& ST_{PC} \,.\, (\overline{ST_{IR_{2,1}}} + \overline{ST_{IR_{2,2}}} + \overline{ST_{IR_{2,3}}} + \overline{ST_{IR_{2,4}}}) \,.\, \overline{XN_{IR_{1,1}}} \,.\, \overline{SQ_{IR_{1,1}}} + XN_{IR_{1,1}} + SQ_{IR_{1,1}} \\
& = (\overline{ST_{IR_{2,1}}} + \overline{ST_{IR_{2,2}}} + \overline{ST_{IR_{2,3}}} + \overline{ST_{IR_{2,4}}}) \,.\, (\overline{ST_{PC}} + ST_{PC}) \,.\, \overline{XN_{IR_{1,1}}} \,.\, \overline{SQ_{IR_{1,1}}} + \\
& (ST_{IR_{2,1}} \,.\, ST_{IR_{2,2}} \,.\, ST_{IR_{2,3}} \,.\, ST_{IR_{2,4}}) \,.\, \overline{XN_{IR_{1,1}}} \,.\, \overline{SQ_{IR_{1,1}}} + XN_{IR_{1,1}} + SQ_{IR_{1,1}} \\
& = (\overline{XN_{IR_{1,1}}} \,.\, \overline{SQ_{IR_{1,1}}}) \,.\, ((\overline{ST_{IR_{2,1}}} + \overline{ST_{IR_{2,2}}} + \overline{ST_{IR_{2,3}}} + \overline{ST_{IR_{2,4}}}) + \\
& (ST_{IR_{2,1}} \,.\, ST_{IR_{2,2}} \,.\, ST_{IR_{2,3}} \,.\, ST_{IR_{2,4}})) + XN_{IR_{1,1}} + SQ_{IR_{1,1}} \\
& = (\overline{XN_{IR_{1,1}}} \,.\, \overline{SQ_{IR_{1,1}}}) + (XN_{IR_{1,1}} + SQ_{IR_{1,1}}) \\
& = 1
\end{aligned}
$$

We have used *Espresso* to minimize the equations. The minimized equations are analyzed to verify whether the properties are violated or not. The complete verification took 41 seconds on a 333 MHz Sun Ultra-5 with 128M RAM. Our framework determined that the Equation (3.33) is violated and generated a simple instruction sequence which violates in-order execution: floating-point addition followed by integer addition. The decode unit issued floating point addition I_{fadd} operation in cycle *n* to floating-point adder pipeline (A1 - A4) and an integer addition operation I_{iadd} to integer ALU (EX) at cycle $n+1$. The instruction I_{iadd} reached join node (MEM unit) prior to I_{fadd}.

We modified the ADL description to change the stall condition depending on current instruction in decode unit and the instructions active in the integer ALU, MUL, FADD, and DIV pipelines. The current instruction will not be issued (decode stalls) if it leads to out-of-order execution. Our framework generated equations for processor model and the properties. The only difference is $ST_{IR_{i,j}}^{self}$ for decode unit (Equation (3.8)) becomes:

$$ST^{self}_{IR_{i,j}} = I_1 \, . \, (S_{2,1} + S_{3,1} + S_{4,1}) + I_2 \, . \, S_{4,2} + I_3 \, . \, (S_{2,3} + S_{4,3})$$

where, the numbers 1, 2, 3, and 4 correspond to the integer ALU, MUL, FADD and DIV pipelines respectively. The signal $S_{x,y}$ is 1 if the latest instruction in pipeline x is active for less than $(\tau(x) - \tau(y))$ cycles. Here, $\tau(x)$ returns the total number of clock cycles needed by pipeline x ($\tau(1) = 1, \tau(2) = 7$, $\tau(3) = 4$, $\tau(4) = 25$). The instructions I_1, I_2, I_3, and I_4 represent the instructions supported by the pipelines 1, 2, 3, and 4 respectively. For example, if current instruction is I_2 (multiply) and there is a instruction in DIV unit which is active for less than 18 cycles ($\tau(4) - \tau(2) = 25 - 7 = 18$), then the decode should stall. Otherwise, it leads to out-of-order execution. Note that, the equation does not have any term for I_4. This is because $S_{x,4}$ can never be 1 since $(\tau(x) - \tau(4))$ is always negative. For the same reason, all the components in the equation does not have four $S_{x,y}$ terms.

The Equation (3.34) is violated for this modeling for $IR_{9,1}$. The instruction sequence generated by our framework for this failure consists of a multiply operation (issued by decode unit in cycle n) followed by a floating-point add operation (issued by decode unit in cycle $(n + 3)$). As a result both the operations reach $IR_{9,1}$ at cycle $(n+7)$. We modified the ADL description to redefine $S_{x,y}$ signal: it is 1 if the latest instruction in pipeline x is active for less than *or equal to* $(\tau(x) - \tau(y))$ cycles. The in-order execution was successful for this modeling. In such a simple situation this kind of specification mistakes might appear as trivial, but when the architecture gets complicated and exploration iterations and varieties increase, the potential for introducing bugs also increases.

E

DESIGN SPACE EXPLORATION

An architect needs to explore the possible design alternatives and consider the application scenarios before finalizing the design decisions. Each design alternative needs to be prototyped and evaluated under typical user environment (using application programs or synthetic benchmarks) to gather necessary estimates including area, power, and performance values. It may be necessary to generate both simulator and hardware (synthesizable HDL) models. The simulator produces profiling data and thus may answer questions concerning the instruction-set, the performance of an algorithm and the required size of memory and registers. However, the hardware prototype is necessary to estimate the required silicon area, clock frequency, and power consumption.

Manual or semi-automatic generation of prototypes is a time consuming process. This can be done only by a set of skilled designers. Furthermore, the interaction among the different teams, such as specification developers, HDL designers, and simulator developers makes rapid architectural exploration infeasible. As a result, system architects rarely have tools or the time to explore architecture alternatives to find the best possible design. This situation is very expensive in both time and engineering resources, and has a substantial impact on time-to-market. Without automation and a unified development environment, the design process is prone to error and may lead to inconsistencies between hardware and software representations.

Figure E.1 shows our ADL-driven architectural exploration framework. The application programs are compiled using the compiler[1] [4] and simulated using the generated simulator. The feedback (performance and code size) is used to modify the ADL specification of the architecture. Similarly, the generated hardware is used to obtain the area, clock frequency, power and performance estimates. The

[1]The compiler is also generated from the architecture specification.

goal is to find the best possible processor, coprocessor, and memory architecture for the given set of application programs. The techniques for generating simulators and hardware models are described in Chapter 4. Section E.1 presents exploration experiments using the generated simulator. The exploration experiments using generated hardware models are described in Section E.2.

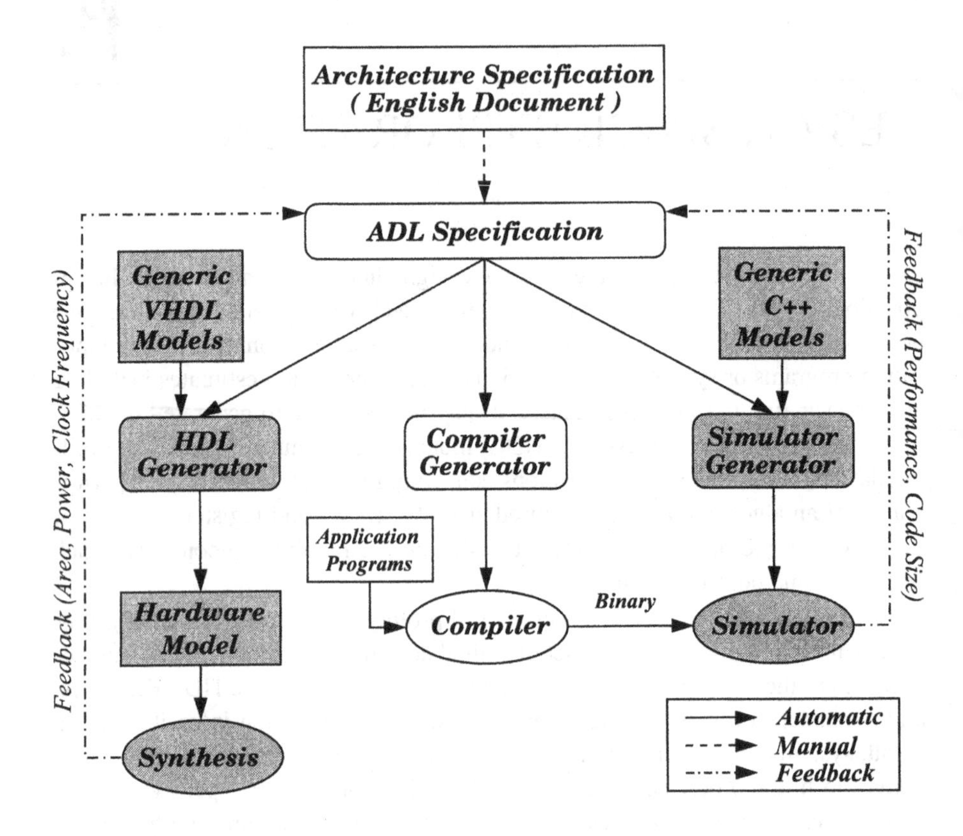

Figure E.1: Architecture exploration framework

E.1 Simulator Generation and Exploration

We have performed extensive exploration experiments by varying different architectural features: exploration varying MIPS R10K processor features [104], coprocessor based exploration [94], and memory subsystem exploration using TI C6x architecture [98].

Exploration varying Processor Features

Contemporary superscalar processors use in-order completion (graduation) to ensure sequential execution behavior in the presence of out-of-order execution. Here, we explore the MIPS R10K processor in the presence of out-of-order graduation without violating functional correctness. The MIPS R10000 [50] is a dynamic, superscalar microprocessor that implements the 64-bit Mips-4 instruction-set architecture. It fetches and decodes four instructions per cycle and dynamically issues them to five fully-pipelined, low-latency execution units. Instructions can be fetched and executed speculatively beyond branches. Instructions graduate in order upon completion.

We have described the MIPS R10K architecture using functional abstraction. The fetch unit function is invoked with the appropriate parameter values. For example, the number of instructions fetched per cycle and the number of instructions sent to decode stage per cycle are set to four. The decode functionality is instantiated with read connection from fetch latch and write connections to MemIssue, IntIssue and FloatIssue units. The decoded instruction is added in the completion queue (ActiveList) which maintains the program order. The decode logic decides where to dispatch (MemIssue, IntIssue or FloatIssue) a particular instruction based on the supported opcodes information available in the control table.

The IntIssue, FloatIssue and MemIssue functions are instantiated with a reservation station size of 16 entries. Each issue unit performs operand read and RAW hazard detection (using appropriate sub-functions) before performing out-of-order issue. Execution units are instantiated using appropriate opcode functionalities. The Address Queue (MemIssue unit) reads data and tag using virtual address while the physical address is computed. It checks whether the load is a hit or miss once the physical address is available. This is different from conventional way of hit or miss detection. In conventional architectures the load request is done using physical address and hit or miss detection is done inside the memory subsystem. This illustrates our ability to reuse the hit or miss detection sub-functions in the processor side (that remains conventionally in the memory side). Similarly, we capture register files and memory hierarchy by instantiating components using appropriate parameters.

We have described the MIPS R10K architecture (with in-order graduation and 8 entry Active List) and generated the software toolkit. We have modified the description to perform out-of-order graduation and generated the simulator. We have used a set of benchmarks from the multimedia and DSP domains.

Figure E.2 presents a subset of the experiments we ran to study the performance improvement due to out-of-order graduation. The light bar presents the number of execution cycles when in-order graduation is used whereas the dark bar presents the number of execution cycles when out-of-order graduation is used. We observe an average performance improvement of 10%. During in-order graduation certain instructions (independent of the instructions above in the Active List) complete execution but are not allowed to graduate since some long latency operations are on top of the Active List and yet to complete. As a result, the Active List becomes full soon and the decode stalls. This situation becomes more prominent when the top instruction is a load and the load misses. We have modified the memory subsystem to study the impact of cache misses along with out-of-order graduation and observed up to 27% performance improvement (in benchmark StateExcerpt when hit ratio is zero). The complete study of the out-of-order graduation for the MIPS R10K processor can be found in [104].

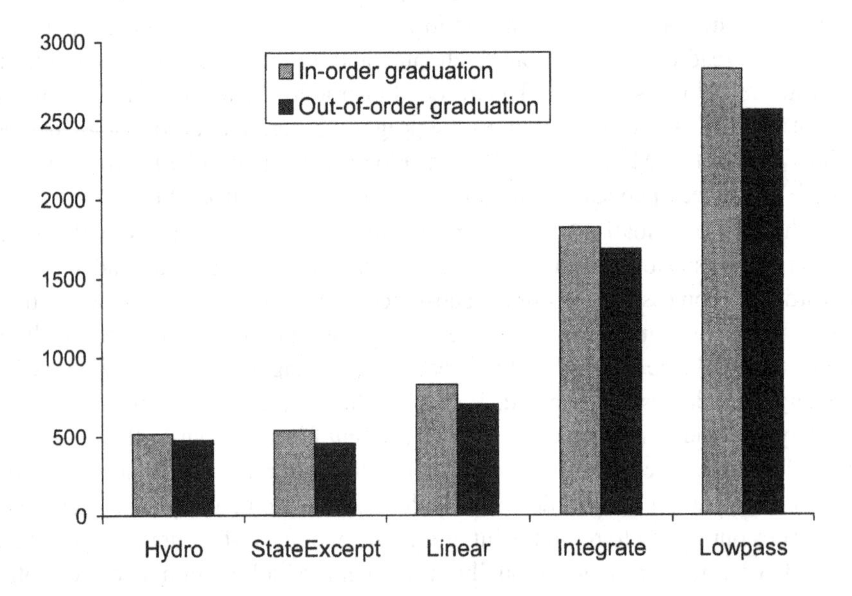

Figure E.2: Cycle counts for different graduation styles

Due to the high modeling efficiency of functional abstraction, the original description and toolkit generation took less than a week; the graduation style modification and toolkit generation took less than a day; the experiments and analysis took few hours; the complete exploration experiment took approximately one week.

Co-processor based Exploration

In the context of co-processor codesign for programmable architectures we have explored the performance impact using a co-processor for the TI C6x environment. TI C6x [131] is an 8-way VLIW DSP processor with a novel memory subsystem (cache hierarchy, configurable SRAM, partitioned register file). TI C6x processor has a deep pipeline, composed of 4 fetch stages (PG, PS, PR, PW), 2 decode stages (DP, DC), followed by the 8 functional units.

We have described the TI C6x architecture using functional abstraction. The fetch functionality consists of four stages viz., program address generation, address send, wait, and receive. Each of the four stages is modeled using respective sub-functions with appropriate parameters. The architecture fetches one VLIW instruction (eight parallel operations) per cycle. The decode function decodes the VLIW word and dispatches up to eight operations per cycle to eight execution units. Each execution unit performs operand read and hazard checks (using sub-functions). At the end of computation each execution unit writes back (using sub-functions) the result to register file.

The functional units, L1, S1, M1 and D1 are connected to the "A" part of the partitioned register file whereas the remaining functional units viz., L2, S2, M2, D2 are connected to the "B" of the register file. Two cross paths, viz., 1X and 2X, are used for transferring data from the other part of the partitioned register file. Each register file is instantiated using generic register file with 16 32-bit registers. Similarly, the memory subsystem consisting of scratch-pad SRAM and cache hierarchy is captured by instantiating components using appropriate parameters.

We have described the TI C6x architecture (where multiplication is done in the functional unit) using functional abstraction and generated the software toolkit. We have modified the description by adding a co-processor (with DMA controller and local memory) that supports multiplication and generated the simulator. This co-processor has its own local memory and uses DMA to transfer data from main memory. We then used a set of DSPStone fixed point benchmarks to explore and evaluate the effects of adding a coprocessor.

Figure E.3 presents a subset of the experiments we ran to study the performance improvement due to the co-processor. The light bar presents the number of execution cycles when the functional unit is used for the multiplication whereas the dark bar presents the number of execution cycles when the co-processor is used. We observe an average performance improvement of 22%. The performance improvement is due to the fact that the co-processor is able to exploit the vector multiplications available in these benchmarks using its local memory. Moreover, functional units operate in register-to-register mode whereas co-processor operates in memory-memory mode. As a result the register pressure and thereby spilling gets reduced in the presence of the co-processor. However, the functional unit performs better when there are mostly scalar multiplications. The complete study of the co-processor based design space exploration can be found in [94].

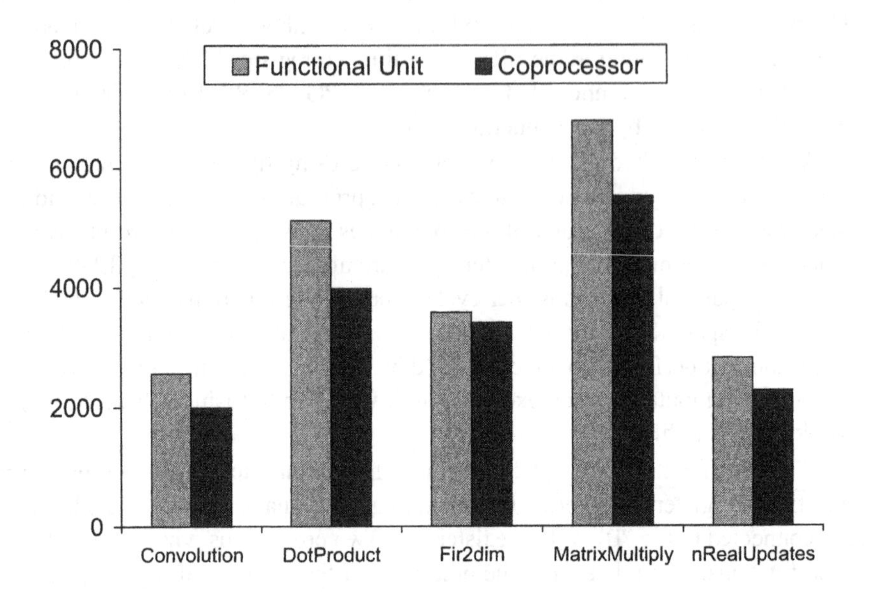

Figure E.3: Functional unit versus coprocessor

Memory Subsystem Exploration

Another important dimension for architectural exploration is the investigation of different memory configurations for a programmable architecture. We explored different memory configurations for the TI C6x architecture with the goal of studying the trade-off between cost and performance. We used a set of benchmarks from the multimedia and DSP domains.

The configurations we experimented with are presented in Table E.1. The numbers in Table E.1 represent: the size of the memory module, the cache/stream buffer organizations: $num\_lines \times num\_ways \times line\_size \times word\_size$, the latency (in number of processor cycles), and the replacement policy (LRU or FIFO). Note that for the stream buffer, num_ways represents the number of FIFO queues present. **The configurations are presented in the increasing order of cost in terms of area.** The first configuration contains an L1 cache and a small stream buffer (256 bytes) to capitalize on the stream nature of the benchmarks. The second configuration contains the L1 cache and an on-chip direct mapped SRAM of 2K. A part of the arrays in the application are mapped to the SRAM. Due to the reduced control necessary for the SRAM, it has a small latency (of 1 cycle), and the area requirements are small. The third configuration contains L1 and L2 caches with FIFO replacement policy. Due to the control necessary for the L2 cache (of size 2K), the cost of this configuration is larger than the configuration 2. Configuration 4 contains an L1 cache, an L2 cache of size 1K and a direct mapped SRAM of size 1K. Due to the extra busses to route the data to the caches and SRAM, this configuration has a larger cost than the previous one. The last configuration contains a large SRAM and has the largest area requirement. All the configurations contain the same off-chip DRAM module with a latency of 20 cycles.

Table E.1: The Memory Subsystem Configurations

Config	L1 Cache	L2 Cache	SRAM	Stream Buffer	DRAM
1	4x2x4x4 latency=1 (LRU)	-	-	4x4x4x4 latency=4	latency=20
2	4x2x4x4 latency=1 (LRU)	-	2K latency=1	-	latency=20
3	4x2x4x4 latency=1 (FIFO)	16x4x8x4 latency=4 (FIFO)	-	-	latency=20
4	4x2x4x4 latency=1 (FIFO)	32x1x8x4 latency=4 (FIFO)	1K latency=1	-	latency=20
5	-	-	8K latency=1	-	latency=20

Figure E.4 presents a subset of the experiments we ran, showing the total cycle counts (including the time spent in the processor) for the set of benchmarks for different memory configurations. Even though these benchmarks are kernels, we observed a significant variation in the trends shown by the different applications.

For instance, in tridiag and stateeq, the first configuration (even though has the lowest cost) performs better (lower cycle count means higher performance), due to the capability of the stream buffer to exploit the stream nature of the access patterns. Moreover, in these applications the most expensive configuration (configuration 5), containing the large SRAM behaves poorly, due to the fact that not all the arrays fit in the SRAM, and the lack of L1 cache to compensate the large latency of the DRAM creates its toll on the performance.

The expected trend of higher cost - higher performance was apparent in the applications integrate and lowpass, While the stream buffer in configuration 1 has a comparable performance to the other configurations, the configuration 5 has the best behavior due to the low latency of the direct mapped on-chip SRAM. The complete study of the memory subsystem exploration can be found in [98].

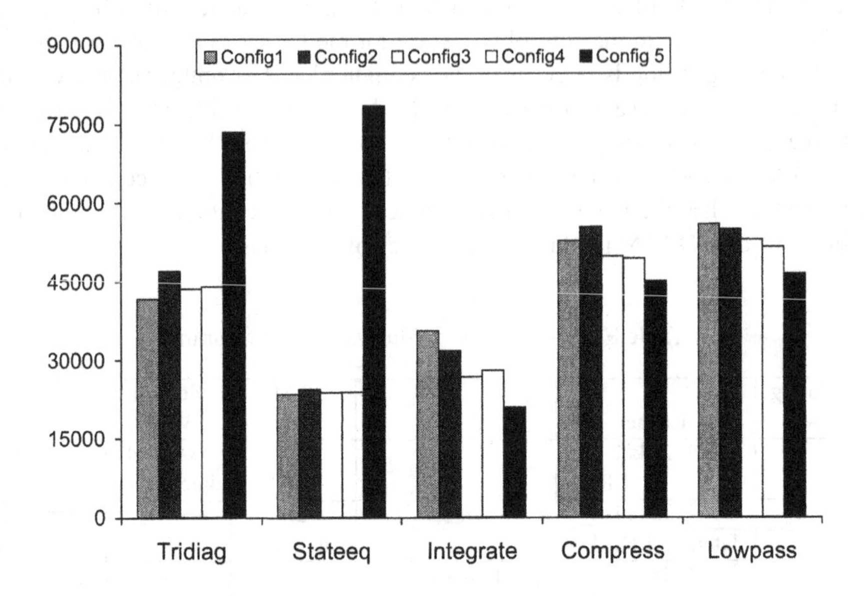

Figure E.4: Cycle counts for the memory configurations

E.2 Hardware Generation and Exploration

We have performed various exploration experiments using the generated hardware models for DLX processor based on silicon area, power, and clock frequency [92, 93]. We have used Synopsys Design Compiler [127] to synthesize the generated HDL description using LSI 10K technology libraries and obtained area, power and clock frequency values.

Table E.2: Synthesis Results: RISC-DLX vs Public-DLX

	HDL Code (lines)	Area (gates)	Speed (MHz)	Power (mW)
RISC-DLX	7758	208 K	35	32.6
Public-DLX	6529	159 K	44	27.4

To ensure the functional correctness, the generated HDL model is validated against the generated simulator using Livermoore loops (LL1 - LL24) and multimedia kernels (compress, GSR, laplace, linear, lowpass, SOR and wavelet). To ensure the fidelity of the generated area, power, and performance numbers, we have compared our generated HDL (RISC version of the DLX) with the hand-written HDL model publicly available from *eda.org* [44]. Table E.2 presents the comparative results between the generated DLX model (*RISC-DLX*) and the hand written DLX model (*Public-DLX*). Our generated design (*RISC-DLX*) is 20-30% off in terms of area, power and clock speed. We believe these are reasonable ranges for early rapid system prototyping and exploration.

```
arg = th2 * piovn
c1 = cos(arg)
s1 = sin(arg)                    int4 = in * 4;
c2 = c1 * c1 - s1 * s1;          j0 = jr * int4 + 1;
s2 = c1 * s1 + c1 * s1;          k0 = ji * int4 + 1;
c3 = c1 * c2 - s1 * s2;
s3 = c2 * s1 + s2 * c1;          jlast = j0 + in - 1;
```

Figure E.5: The application program

Figure E.5 shows one of the most frequently executed code segment from FFT benchmark that we have used as an application program during micro-architectural exploration.

In this section we present three exploration experiments: pipeline path exploration, pipeline stage exploration and instruction-set exploration. The reported area, power, and clock frequency numbers are for the execution units only. The numbers do not include the contributions from others components such as *Fetch*, *Decode*, *MEM* and *WriteBack*.

Addition of Functional Units (Pipeline Paths)

Figure E.6 shows the exploration results due to addition of pipeline paths using the application program shown in Figure E.5. The first configuration has only one pipeline path consisting of *Fetch*, *Decode*, one execution unit (*Ex1*), *MEM* and *WriteBack*. The *Ex1* unit supports five operations: *sin*, *cos*, +, - and ×. The second configuration is exactly same as the first configuration except it has one more execution unit (*Ex2*) parallel to *Ex1*. The *Ex2* unit supports three operations: +, - and ×. Similarly, the third configuration has three parallel execution units: *Ex1* (+, -, ×), *Ex2* (+, -, ×) and *Ex3* (*sin*, *cos*, +, - and ×). Finally, the fourth configuration has four parallel execution units: *Ex1* (sin, cos), *Ex2* (+, -, MAC[2]),

[2]MAC performs multiply-and-accumulate of the form $a \times b + c$

Ex3 and *Ex4*, where *Ex3* and *Ex4* are customized functional units that perform $a \times b + c \times d$.

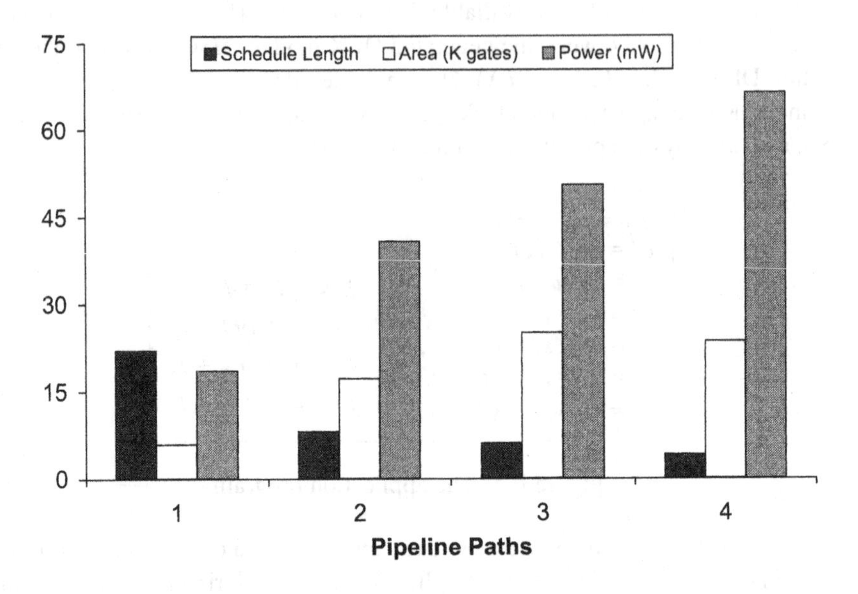

Figure E.6: Pipeline path exploration

The application program requires fewer number of cycles (schedule length) due to the addition of pipeline paths whereas the area and power requirement increases. The fourth configuration is interesting since both area and schedule length decrease due to addition of specialized hardware and removal of operations from other execution units.

Addition of Pipeline Stages

Figure E.7 presents exploration experiments due to addition of pipeline stages in the multiplier unit. The first configuration is a one-stage multi-cycle multiplier. The second, third and fourth configurations use multipliers with two, three and four stages respectively. The clock frequency (speed) is improved due to addition of pipeline stages. The fourth configuration generated 30% speed improvement at the cost of 13% area increase over the third configuration.

Addition of Operations

Figure E.8 presents exploration results for addition of opcodes using three processor configurations. The three configurations are shown in Figure E.8. The first configuration has four parallel execution units: *FU1*, *FU2*, *FU3* and *FU4*. The *FU1* supports three operations: +, -, and ×. The *FU2*, *FU3* and *FU4* supports (+, -, ×), (*and*, *or*) and (*sin*, *cos*) respectively. The second configuration is obtained by adding a *cos* operation in the *FU3* of the first configuration. This generated reduction of schedule length of the application program at the cost of increase in area. The third configuration is obtained by adding multipliers both in *FU3* and *FU4* of the second configuration. This generated best possible (using +, -, ×, *sin* and *cos*) schedule length for the application program shown in Figure E.5.

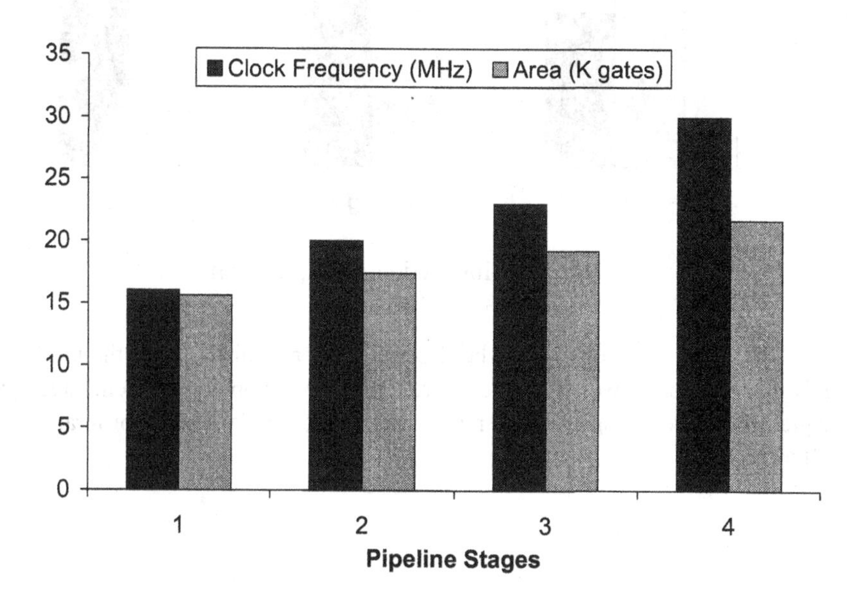

Figure E.7: Pipeline stage exploration

Each iteration in our exploration framework is in the order of hours to days depending on the amount of modification needed in the ADL and the synthesis time. However, each iteration will be in the order of weeks to months for manual or semi-automatic development of HDL models. The reduction of exploration time is at least an order of magnitude.

We have also performed various micro-architectural explorations of the MIPS 4000 processor. A public release of the exploration framework is available from http://www.cecs.uci.edu/~express. This release also supports graphical user inter-

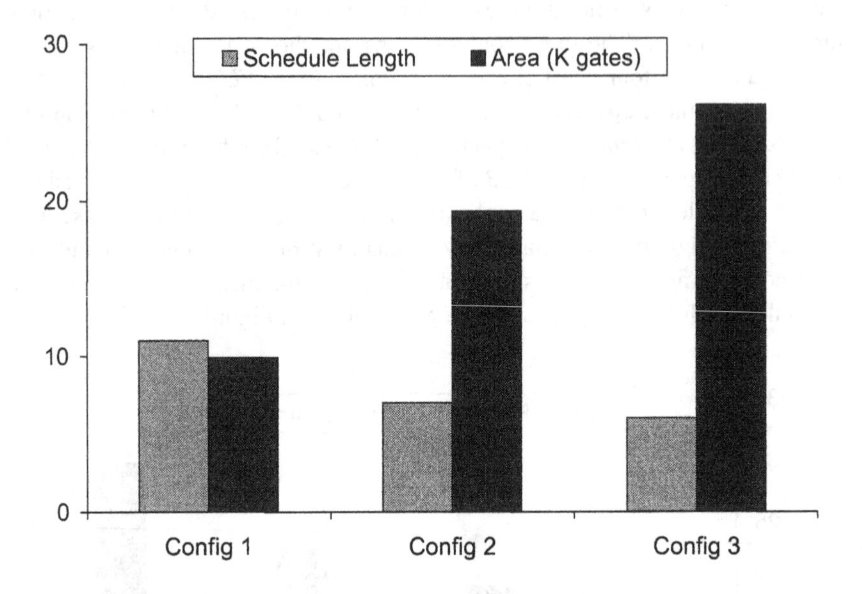

Figure E.8: Instruction-set exploration

face (GUI). The architecture can be described (or modified) using the GUI. The ADL specification as well as the software toolkit are automatically generated from the graphical description to enable rapid design space exploration of programmable architectures.

References

[1] A. Aharon and D. Goodman and M. Levinger and Y. Lichtenstein and Y. Malka and C. Metzger and M. Molcho and G. Shurek. Test program generation for functional verification of PowerPC processors in IBM. In *Proceedings of Design Automation Conference (DAC)*, pages 279–285, 1995.

[2] A. Bunker and G. Gopalakrishnan and S. Mckee. Validation, verification, and testing of computer software. *ACM Computing Surveys*, 9(1):1–32, January 2004.

[3] A. Fauth and A. Knoll. Automatic generation of DSP program development tools. In *Proceedings of Int'l Conf. Acoustics, Speech and Signal Processing (ICASSP)*, pages 457–460, 1993.

[4] A. Halambi and A. Shrivastava and N. Dutt and A. Nicolau. A customizable compiler framework for embedded systems. In *Proceedings of Software and Compilers for Embedded Systems (SCOPES)*, 2001.

[5] A. Halambi and P. Grun and V. Ganesh and A. Khare and N. Dutt and A. Nicolau. EXPRESSION: A language for architecture exploration through compiler/simulator retargetability. In *Proceedings of Design Automation and Test in Europe (DATE)*, pages 485–490, 1999.

[6] A. Inoue and H. Tomiyama and F. Eko and H. Kanbara and H. Yasuura. A programming language for processor based embedded systems. In *Proceedings of Asia Pacific Conference on Hardware Description Languages (APCHDL)*, pages 89–94, 1998.

[7] A. Inoue and H. Tomiyama and H. Okuma and H. Kanbara and H. Yasuura. Language and compiler for optimizing datapath widths of embedded systems. *IEICE Trans. Fundamentals*, E81-A(12):2595–2604, 1998.

[8] A. Khare and N. Savoiu and A. Halambi and P. Grun and N. Dutt and A. Nicolau. V-SAT: A visual specification and analysis tool for system-on-chip exploration. In *Proceedings of EUROMICRO Conference*, pages 1196–1203, 1999.

[9] A. Nohl and G. Braun and O. Schliebusch and R. Leupers and H. Meyr and A. Hoffmann. A universal technique for fast and flexible instruction-set architecture simulation. In *Proceedings of Design Automation Conference (DAC)*, pages 22–27, 2002.

[10] ARC Cores. *http://www.arccores.com.*

[11] Bob Bentley. High level validation of next-generation microprocessors. In *Proceedings of High Level Design Validation and Test (HLDVT)*, 2002.

[12] C. Ejik. Sequential equivalence checking without state space traversal. In *Proceedings of Design Automation and Test in Europe (DATE)*, pages 618–623, 1998.

[13] C. Jacobi. Formal verification of complex out-of-order pipelines by combining model-checking and theorem-proving. In E. Brinksma and K. Larsen, editor, *Proceedings of Computer Aided Verification (CAV)*, volume 2404 of *LNCS*, pages 309–323. Springer-Verlag, 2002.

[14] C. Seger and R. Bryant. Formal verification by symbolic evaluation of partially-ordered trajectories. In *Formal Methods in System Design*, volume 6, pages 147–189, March 1995.

[15] C. Siska. A processor description language supporting retargetable multi-pipeline DSP program development tools. In *Proceedings of International Symposium on System Synthesis (ISSS)*, pages 31–36, 1998.

[16] Chris Basoglu and Woobin Lee and John Setel O'Donnell. *The MAP1000A VLIW Mediaprocessor*, 2000.

[17] CoWare LISATek Products. *http://www.coware.com.*

[18] D. Anastasakis and R. Damiano and H. Ma and T. Stanion. A practical and efficient method for compare-point matching. In *Proceedings of Design Automation Conference (DAC)*, pages 305–310, 2002.

[19] D. Campenhout and T. Mudge and J. Hayes. High-level test generation for design verification of pipelined microprocessors. In *Proceedings of Design Automation Conference (DAC)*, pages 185–188, 1999.

[20] D. Cyrluk. Microprocessor verification in PVS: A methodology and simple example. Technical report, SRI-CSL-93-12, 1993.

[21] D. Kastner. TDL: A hardware and assembly description languages. Technical Report TDL 1.4, Saarland University, Germany, 2000.

[22] D. Lanneer and J. Praet and A. Kifli and K. Schoofs and W. Geurts and F. Thoen and G. Goossens. CHESS: Retargetable code generation for embedded DSP processors. In *Code Generation for Embedded Processors.*, pages 85–102. Kluwer Academic Publishers, 1995.

[23] E. Schnarr and J. Larus. Fast out-of-order processor simulation using memoization. In *Architectural Support for Programming Languages and Operating Systems (ASPLOS)*, pages 283–294, 1998.

[24] E. Schnarr and M. Hill and J. Larus. Facile: A language and compiler for high-performance processor simulators. In *Programming Language Design and Implementation (PLDI)*, pages 321–331, 2001.

[25] E. Witchel and M. Rosenblum. Embra: Fast and flexible machine simulation. In *Measurement and Modeling of Computer Systems*, pages 68–79, 1996.

[26] F. Corno and G. Cumani and M. Reorda and G. Squillero. Fully automatic test program generation for microprocessor cores. In *Proceedings of Design Automation and Test in Europe (DATE)*, pages 1006–1011, 2003.

[27] F. Engel and J. Nuhrenberg and G. Fettweis. A generic tool set for application specific processor architectures. In *Proceedings of International Symposium on Hardware/Software Codesign (CODES)*, 2000.

[28] F. Lohr and A. Fauth and M. Freericks. Sigh/sim: An environment for retargetable instruction set simulation. Technical Report 1993/43, Dept. Computer Science, Tech. Univ. Berlin, Germany, 1993.

[29] F. Pong and M. Dubois. Verification techniques for cache coherence protocols. *ACM Computing Surveys*, 29(1):82–126, 1997.

[30] G. Hadjiyiannis and P. Russo and S. Devadas. A methodology for accurate performance evaluation in architecture exploration. In *Proceedings of Design Automation Conference (DAC)*, pages 927–932, 1999.

[31] G. Hadjiyiannis and S. Hanono and S. Devadas. ISDL: An instruction set description language for retargetability. In *Proceedings of Design Automation Conference (DAC)*, pages 299–302, 1997.

[32] Gordon Moore. Cramming more components onto integrated circuits. *Electronics*, 38(8), 1965.

[33] Gregory S Spirakis. *Designing for 65nm and Beyond.* Keynote Address at Design Automation and Test in Europe (DATE), 2004.

[34] H. Akaboshi. *A Study on Design Support for Computer Architecture Design.* PhD thesis, Dept. of Information Systems, Kyushu University, Japan, Jan 1996.

[35] H. Akaboshi and H. Yasuura. Behavior extraction of MPU from HDL description. In *Proceedings of Asia Pacific Conference on Hardware Description Languages (APCHDL)*, 1994.

[36] H. Chockler and O. Kupferman and R. Kurshan and M. Vardi. A practical approach to coverage in model checking. In *Proceedings of Computer Aided Verification (CAV)*, volume 2102 of *LNCS*, pages 66–78. Springer-Verlag, 2001.

[37] H. Iwashita and S. Kowatari and T. Nakata and F. Hirose. Automatic test pattern generation for pipelined processors. In *Proceedings of International Conference on Computer-Aided Design (ICCAD)*, pages 580–583, 1994.

[38] H. Tomiyama and A. Halambi and P. Grun and N. Dutt and A. Nicolau. Architecture description languages for systems-on-chip design. In *Proceedings of Asia Pacific Conference on Chip Design Language*, pages 109–116, 1999.

[39] http://pjro.metsa.astem.or.jp/udli. *UDL/I Simulation/Synthesis Environment*, 1997.

[40] http://www-cad.eecs.berkeley.edu/Software/software.html. *Espresso.*

[41] http://www-ee.engr.ccny.cuny.edu/notes/ee210/eqntott_man.html. *Eqntott.*

[42] http://www.axysdesign.com. *Axys Design Automation.*

[43] http://www.cs.cmu.edu/~modelcheck. *Symbolic Model Verifier.*

[44] http://www.eda.org/rassp/vhdl/models/processor.html. *Synthesizable DLX.*

[45] http://www.ics.uci.edu/~express. *Exploration framework using EXPRESSION.*

[46] http://www.improvsys.com. *Improv Inc.*

[47] http://www.lucent.com/micro/Starcore. *Starcore, Next Generation DSPs.*

[48] http://www.motorola.com. *MPC7450 Microprocessor.*

[49] http://www.retarget.com. *Target Compiler Technologies.*

[50] http://www.sgi.com. *MIPS R10000 Microprocessor.*

[51] http://www.simplescalar.com. *Simplescalar.*

[52] http://www.sparc.com/resource.htm#V8. *The SPARC Architecture Manual, Version 8.*

[53] J. Burch and D. Dill. Automatic verification of pipelined microprocessor control. In D. Dill, editor, *Proceedings of Computer Aided Verification (CAV)*, volume 818 of *LNCS*, pages 68–80. Springer-Verlag, 1994.

[54] J. Gyllenhaal and B. Rau and W. Hwu. HMDES version 2.0 specification. Technical Report IMPACT-96-3, IMPACT Research Group, Univ. of Illinois, Urbana. IL, 1996.

[55] J. Hennessy and D. Patterson. *Computer Architecture: A Quantitative Approach.* Morgan Kaufmann Publishers Inc, San Mateo, CA, 1990.

[56] J. Huggins and D. Campenhout. Specification and verification of pipelining in the arm2 risc microprocessor. *ACM Transactions on Design Automation of Electronic Systems (TODAES)*, 3(4):563–580, October 1998.

[57] J. Levitt and K. Olukotun. Verifying correct pipeline implementation for microprocessors. In *Proceedings of International Conference on Computer-Aided Design (ICCAD)*, pages 162–169, 1997.

[58] J. Marques-Silva and T. Glass. Combinational equivalence checking using satisfiability and recursive learning. In *Proceedings of Design Automation and Test in Europe (DATE)*, pages 145–149, 1999.

[59] J. Miyake and G. Brown and M. Ueda and T. Nishiyama. Automatic test generation for functional verification of microprocessors. In *Proceedings of Asian Test Symposium (ATS)*, pages 292–297, 1994.

[60] J. Paakki. Attribute grammar paradigms - a high level methodology in language implementation. *ACM Computing Surveys*, 27(2):196–256, June 1995.

[61] J. Sato and A. Alomary and Y. Honma and T. Nakata and A. Shiomi and N. Hikichi and M. Imai. PEAS-I: A hardware/software codesign systems for ASIP development. *IEICE Trans. Fundamentals*, E77-A(3):483–491, 1994.

[62] J. Sawada and W. D. Hunt. Processor verification with precise exceptions and speculative execution. In A. Hu and M. Vardi, editor, *Proceedings of Computer Aided Verification (CAV)*, volume 1427 of *LNCS*, pages 135–146. Springer-Verlag, 1998.

[63] J. Shen and J. Abraham and D. Baker and T. Hurson and M. Kinkade and G. Gervasio and C. Chu and G. Hu. Functional verification of the equator MAP1000 microprocessor. In *Proceedings of Design Automation Conference (DAC)*, pages 169–174, 1999.

[64] J. Skakkebaek and R. Jones and D. Dill. Formal verification of out-of-order execution using incremental flushing. In A. Hu and M. Vardi, editor, *Proceedings of Computer Aided Verification (CAV)*, volume 1427 of *LNCS*, pages 98–109. Springer-Verlag, 1998.

[65] J. Zhu and D. Gajski. A retargetable, ultra-fast, instruction set simulator. In *Proceedings of Design Automation and Test in Europe (DATE)*, 1999.

[66] K. Kohno and N. Matsumoto. A new verification methodology for complex pipeline behavior. In *Proceedings of Design Automation Conference (DAC)*, pages 816–821, 2001.

[67] Kanna Shimizu. *Writing, Verifying, and Exploiting Formal Specifications for Hardware Designs*. PhD thesis, Stanford University, 2002.

[68] L. Chen and S. Ravi and A. Raghunathan and S. Dey. A scalable software-based self-test methodology for programmable processors. In *Proceedings of Design Automation Conference (DAC)*, pages 548–553, 2003.

[69] L. Wang and M. Abadir and N. Krishnamurthy. Automatic generation of assertions for formal verification of PowerPC microprocessor arrays using symbolic trajectory evaluation. In *Proceedings of Design Automation Conference (DAC)*, pages 534–537, 1998.

[70] LEON2 Processor. *http://www.gaisler.com/leon.html*.

[71] M. Aagaard and B. Cook and N. Day and R. Jones. A framework for microprocessor correctness statements. In T. Margaria and T. Melham, editor, *Proceedings of Correct Hardware Design and Verification Methods (CHARME)*, volume 2144 of *LNCS*, pages 433–448. Springer-Verlag, 2001.

[72] M. Freericks. The nML machine description formalism. Technical Report TR SM-IMP/DIST/08, TU Berlin CS Dept., 1993.

[73] M. Hartoog and J. Rowson and P. Reddy and S. Desai and D. Dunlop and E. Harcourt and N. Khullar. Generation of software tools from processor descriptions for hardware/software codesign. In *Proceedings of Design Automation Conference (DAC)*, pages 303–306, 1997.

[74] M. Hohenauer and H. Scharwaechter and K. Karuri and O. Wahlen and T. Kogel and R. Leupers and G. Ascheid and H. Meyr and G. Braun and H. Someren. A methodology and tool suite for c compiler generation from ADL processor models. In *Proceedings of Design Automation and Test in Europe (DATE)*, pages 1276–1283, 2004.

[75] M. Itoh and S. Higaki and Y. Takeuchi and A. Kitajima and M. Imai and J. Sato and A. Shiomi. PEAS-III: An ASIP design environment. In *Proceedings of International Conference on Computer Design (ICCD)*, 2000.

[76] M. Itoh and Y. Takeuchi and M. Imai and A. Shiomi. Synthesizable HDL generation for pipelined processors from a micro-operation description. *IEICE Trans. Fundamentals*, E00-A(3), March 2000.

[77] M. Reshadi and P. Mishra and N. Dutt. Instruction set compiled simulation: A technique for fast and flexible instruction set simulation. In *Proceedings of Design Automation Conference (DAC)*, pages 758–763, 2003.

[78] M. Srivas and M. Bickford. Formal verification of a pipelined microprocessor. In *IEEE Software*, volume 7(5), pages 52–64, 1990.

[79] M. Velev and R. Bryant. Formal verification of superscalar microprocessors with multicycle functional units, exceptions, and branch prediction. In *Proceedings of Design Automation Conference (DAC)*, pages 112–117, 2000.

[80] The MDES User Manual. *http://www.trimaran.org*, 1997.

[81] MIPS Technologies, Inc. *MIPS R4000 Microprocessor User's Manual*, 1994.

[82] N. Krishnamurthy and M. Abadir and A. Martin and J. Abraham. Design and development paradigm for industrial formal verification tools. *IEEE Design & Test of Computers*, 18(4):26–35, July-August 2001.

[83] N. Medvidovic and R. Taylor. A framework for classifying and comparing architecture description languages. In M. Jazayeri and H. Schauer, editor,

Proceedings of European Software Engineering Conference (ESEC), pages 60–76. Springer–Verlag, 1997.

[84] O. Schliebusch and A. Chattopadhyay and M. Steinert and G. Braun and A. Nohl and R. Leupers and G. Ascheid and H. Meyr. RTL processor synthesis for architecture exploration and implementation. In *Proceedings of Design Automation and Test in Europe (DATE)*, pages 156–160, 2004.

[85] O. Schliebusch and A. Hoffmann and A. Nohl and G. Braun and H. Meyr. Architecture implementation using the machine description language LISA. In *Proceedings of Asia South Pacific Design Automation Conference (ASP-DAC) / International Conference on VLSI Design*, pages 239–244, 2002.

[86] O. Wahlen and M. Hohenauer and R. Leupers and H. Meyr. Instruction scheduler generation for retragetable compilation. *IEEE Design & Test of Computers*, 20(1):34–41, Jan-Feb 2003.

[87] P. Ammann and P. Black and W. Majurski. Using model checking to generate tests from specifications. In *Proceedings of International Conference on Formal Engineering Methods (ICFEM)*, pages 46–54, 1998.

[88] P. Grun and A. Halambi and N. Dutt and A. Nicolau. RTGEN: An algorithm for automatic generation of reservation tables from architectural descriptions. *IEEE Transactions on Very Large Scale Integration (VLSI) Systems*, 11(4):731–737, August 2003.

[89] P. Grun and N. Dutt and A. Nicolau. Memory aware compilation through accurate timing extraction. In *Proceedings of Design Automation Conference (DAC)*, pages 316–321, 2000.

[90] P. Ho and A. Isles and T. Kam. Formal verification of pipeline control using controlled token nets and abstract interpretation. In *Proceedings of International Conference on Computer-Aided Design (ICCAD)*, pages 529–536, 1998.

[91] P. Ho and Y. Hoskote and T. Kam and M. Khaira and J. O'Leary and X. Zhao and Y. Chen and E. Clarke. Verification of a complete floating-point unit using word-level model checking. In M. Srivas and A. Camilleri, editor, *Proceedings of Formal Methods in Computer-Aided Design (FMCAD)*, volume 1166 of *LNCS*, pages 19–33. Springer-Verlag, 1996.

[92] P. Mishra and A. Kejariwal and N. Dutt. Rapid exploration of pipelined processors through automatic generation of synthesizable RTL models. In *Proceedings of Rapid System Prototyping (RSP)*, pages 226–232, 2003.

[93] P. Mishra and A. Kejariwal and N. Dutt. Synthesis-driven exploration of pipelined embedded processors. In *Proceedings of International Conference on VLSI Design*, 2004.

[94] P. Mishra and F. Rousseau and N. Dutt and A. Nicolau. Architecture description language driven design space exploration in the presence of coprocessors. In *Proceedings of Synthesis and System Integration of Mixed Technologies (SASIMI)*, 2001.

[95] P. Mishra and H. Tomiyama and N. Dutt and A. Nicolau. Architecture description language driven verification of in-order execution in pipelined processors. Technical Report UCI-ICS 01-20, University of California, Irvine, 2000.

[96] P. Mishra and H. Tomiyama and N. Dutt and A. Nicolau. Automatic verification of in-order execution in microprocessors with fragmented pipelines and multicycle functional units. In *Proceedings of Design Automation and Test in Europe (DATE)*, pages 36–43, 2002.

[97] P. Mishra and J. Astrom and N. Dutt and A. Nicolau. Functional abstraction of programmable embedded systems. Technical Report UCI-ICS 01-04, University of California, Irvine, January 2001.

[98] P. Mishra and M. Mamidipaka and N. Dutt. Processor-memory coexploration using an architecture description language. *ACM Transactions on Embedded Computing Systems (TECS)*, 3(1):140–162, 2004.

[99] P. Mishra and N. Dutt. Automatic modeling and validation of pipeline specifications. *ACM Transactions on Embedded Computing Systems (TECS)*, 3(1):114–139, 2004.

[100] P. Mishra and N. Dutt. Graph-based functional test program generation for pipelined processors. In *Proceedings of Design Automation and Test in Europe (DATE)*, pages 182–187, 2004.

[101] P. Mishra and N. Dutt. Architecture description languages for programmable embedded systems. *IEE Proceedings on Computers and Digital Techniques*, 2005.

[102] P. Mishra and N. Dutt and A. Nicolau. Functional abstraction driven design space exploration of heterogeneous programmable architectures. In *Proceedings of International Symposium on System Synthesis (ISSS)*, pages 256–261, 2001.

[103] P. Mishra and N. Dutt and A. Nicolau. Specification of hazards, stalls, interrupts, and exceptions in EXPRESSION. Technical Report UCI-ICS 01-05, University of California, Irvine, 2001.

[104] P. Mishra and N. Dutt and A. Nicolau. A study of out-of-order completion for the MIPS R10K superscalar processor. Technical Report UCI-ICS 01-06, University of California, Irvine, January 2001.

[105] P. Mishra and N. Dutt and H. Tomiyama. Towards automatic validation of dynamic behavior in pipelined processor specifications. *Kluwer Design Automation for Embedded Systems(DAES)*, 8(2-3):249–265, June-September 2003.

[106] P. Mishra and N. Dutt and N. Krishnamurthy and M. Abadir. A top-down methodology for validation of microprocessors. *IEEE Design & Test of Computers*, 21(2):122–131, 2004.

[107] P. Mishra and P. Grun and N. Dutt and A. Nicolau. Processor-memory co-exploration driven by an architectural description language. In *Proceedings of International Conference on VLSI Design*, pages 70–75, 2001.

[108] P. Paulin and C. Liem and T. May and S. Sutarwala. FlexWare: A flexible firmware development environment for embedded systems. In *Prof. of Dagstuhl Workshop on Code Generation for Embedded Processors*, pages 67–84, 1994.

[109] Paul C. Clements. A survey of architecture description languages. In *Proceedings of International Workshop on Software Specification and Design (IWSSD)*, pages 16–25, 1996.

[110] Prabhat Mishra. *Specification-driven Validation of Programmable Embedded Systems*. PhD thesis, University of California, Irvine, March 2004.

[111] R. Bryant. Symbolic simulation - techniques and applications. In *Proceedings of Design Automation Conference (DAC)*, pages 517–521, 1990.

[112] R. Bryant and C. Seger. Formal verification of digital circuits using symbolic ternary system models. In *Proceedings of Computer Aided Verification (CAV)*, pages 121–146, 1990.

[113] R. Cmelik and D. Keppel. Shade: A fast instruction-set simulator for execution profiling. *ACM SIGMETRICS Performance Evaluation Review*, 22(1):128–137, May 1994.

[114] R. Ho and C. Yang and Mark A. Horowitz and D. Dill. Architecture validation for processors. In *Proceedings of International Symposium on Computer Architecture (ISCA)*, 1995.

[115] R. Jhala and K. L. McMillan. Microarchitecture verification by compositional model checking. In G. Berry et al., editor, *Proceedings of Computer Aided Verification (CAV)*, volume 2102 of *LNCS*, pages 396–410. Springer-Verlag, 2001.

[116] R. Leupers and P. Marwedel. Retargetable generation of code selectors from HDL processor models. In *Proceedings of European Design and Test Conference (EDTC)*, pages 140–144, 1997.

[117] R. Leupers and P. Marwedel. Retargetable code generation based on structural processor descriptions. *Design Automation for Embedded Systems*, 3(1):75–108, 1998.

[118] R. M. Hosabettu. *Systematic Verification Of Pipelined Microprocessors*. PhD thesis, Department of Computer Science, University of Utah, 2000.

[119] S. Fine and A. Ziv. Coverage directed test generation for functional verification using bayesian networks. In *Proceedings of Design Automation Conference (DAC)*, pages 286–291, 2003.

[120] S. Hanono and S. Devadas. Instruction selection, resource allocation, and scheduling in the AVIV retargetable code generator. In *Proceedings of Design Automation Conference (DAC)*, pages 510–515, 1998.

[121] S. Pees and A. Hoffmann and H. Meyr. Retargetable compiled simulation of embedded processors using a machine description language. *ACM Transactions on Design Automation of Electronic Systems*, 5(4):815–834, Oct. 2000.

[122] S. Thatte and J. Abraham. Test generation for microprocessors. *IEEE Transactions on Computers*, C-29(6):429–441, June 1980.

[123] S. Ur and Y. Yadin. Micro architecture coverage directed generation of test programs. In *Proceedings of Design Automation Conference (DAC)*, pages 175–180, 1999.

[124] S. Wang and S. Malik. Synthesizing operating system based device drivers in embedded systems. In *Proceedings of International Symposium on Hardware/Software Codesign and System Synthesis (CODES+ISSS)*, pages 37–44, 2003.

[125] StrongArm. *StrongARM Processors.* http://developer.intel.com, 2000.

[126] SUN Microsystems. *UltraSPARC IIi User's Manual*, 1997.

[127] Synopsys. *http://www.synopsys.com.*

[128] Synopsys Formality. *http://www.synopsys.com.*

[129] T. Morimoto and K. Yamazaki and H. Nakamura and T. Boku and K. Nakazawa. Superscalar processor design with hardware description language aidl. In *Proceedings of Asia Pacific Conference on Hardware Description Languages (APCHDL)*, 1994.

[130] Tensilica Inc. *http://www.tensilica.com.*

[131] Texas Instruments. *TMS320C6201 CPU and Instruction Set Reference Guide*, 1998.

[132] V. Rajesh and Rajat Moona. Processor modeling for hardware software codesign. In *Proceedings of International Conference on VLSI Design*, pages 132–137, 1999.

[133] V. Zivojnovic and S. Pees and H. Meyr. LISA - machine description language and generic machine model for HW/SW co-design. In *IEEE Workshop on VLSI Signal Processing*, pages 127–136, 1996.

[134] Verisity Design, Inc. *http://www.verisity.com.*

[135] Verisity Verification Vault. *https://www.verificationvault.com.*

[136] W. Adrion and M. Branstad and J. Cherniavsky. Validation, verification, and testing of computer software. *ACM Computing Surveys*, 14(2):159–192, June 1982.

[137] W. Qin and S. Malik. *Architecture Description Languages for Retargetable Compilation, in The Compiler Design Handbook: Optimizations & Machine Code Generation.* CRC Press, 2002.

[138] www.intel.com. *IA-64 Architecture.*

[139] www.rs.e-technik.tu-darmstadt.de/TUD/res/dlxdocu/SuperscalarDLX.html. *A Superscalar Version of the DLX Processor.*

Index

ADL, 10, 16
antecedent, 86
application programs, 3
architectural exploration, 155
architectural flaws, 8
Architecture Description Language, 16
architecture manual, 15
architecture validation, 83
assembler, 16
attribute grammar, 18

BDD, 85
behavior, 16, 25
behavioral ADL, 18, 130
Boolean, 83
branch taken, 52
bubble insertion, 49

code coverage, 8
compiler, 16
completeness, 10, 39
connectedness, 10, 34
consequent, 86
constrained-random, 113
coprocessor, 8, 15
counterexample, 98

data-transfer edge, 31
data-transfer path, 24, 31
debugger, 16
design complexity, 3
design space exploration, 16
determinism, 10, 55, 151

DMA, 23
DSE, 16
DSP, 11
dynamic behavior, 48

embedded systems, 3
equivalence checking, 10, 87
exception, 147
execution edge, 33
execution path, 103
exploration, 16
EXPRESSION, 17

false data-transfer path, 37
false pipeline path, 35
fault model, 103
finiteness, 10, 41
formal techniques, 83
formal verification, 9, 83
FSM, 8
functional abstraction, 11, 69
functional coverage, 7, 95, 103, 105
functional errors, 4
functional verification, 4

golden reference model, 7
graph coverage, 98
graph model, 96

hardware ADL, 16
hardware description language, 21
HDL, 19

implementation, 83

implementation bugs, 8
in-order execution, 10, 55, 151
instruction register, 51
instruction-set, 16
instruction-set simulation, 9
interface synthesis, 18
interrupt, 147
interrupt handler, 149
ISDL, 17

LISA, 17
logic bugs, 4
LRU, 88

MDES, 17
memory, 15
memory subsystem, 8
MIMOLA, 17
mixed ADL, 18, 134
MMU, 88
model checker, 96
modeling language, 20

nML, 17
normal flow, 49

operation edge, 33
operation execution, 103

partial ADL, 18
pipeline behavior, 29
pipeline edge, 31
pipeline execution, 103
pipeline latch, 51
pipeline path, 24, 31
processor, 15
processor core, 8
programmable architectures, 3
programmable components, 3
programming language, 20
property checking, 83

RADL, 17
reference model, 77
register read/write, 103
register-transfer level, 5
RISC, 11
RTL, 5
RTL design, 83

sequential execution, 52
Sim-nML, 17
simulation, 83
simulation vectors, 5
simulator, 16
SMV, 57
SOC, 4
Software ADL, 16
software toolkit, 16
specification, 15
specification language, 15
squash, 49
stall, 49
state space explosion, 83
static behavior, 30
static verification, 87
STE, 85
structural ADL, 19, 127
structure, 16, 24, 31
superscalar, 11
symbolic simulation, 85
synthesis, 16
system-on-chip, 3

ternary simulation, 85
test generation, 16, 95, 106
test programs, 99
top-down validation, 8

Valen-C, 17
validation, 16
VLIW, 11